Hands-On Programming with R

Garrett Grolemund

Beijing · Cambridge · Farnham · Köln · Sebastopol · Tokyo

Hands-On Programming with R

by Garrett Grolemund

Copyright © 2014 Garrett Grolemund. All rights reserved.

Printed in the United States of America.

Published by O'Reilly Media, Inc., 1005 Gravenstein Highway North, Sebastopol, CA 95472.

O'Reilly books may be purchased for educational, business, or sales promotional use. Online editions are also available for most titles (*http://safaribooksonline.com*). For more information, contact our corporate/institutional sales department: 800-998-9938 or *corporate@oreilly.com*.

Editors: Julie Steele and Courtney Nash
Production Editor: Matthew Hacker
Copyeditor: Eliahu Sussman
Proofreader: Amanda Kersey

Indexer: Judith McConville
Cover Designer: Randy Comer
Interior Designer: David Futato
Illustrator: Rebecca Demarest

July 2014:   First Edition

Revision History for the First Edition:

2014-07-08:   First release

2015-01-23:   Second release

See *http://oreilly.com/catalog/errata.csp?isbn=9781449359010* for release details.

Nutshell Handbook, the Nutshell Handbook logo, and the O'Reilly logo are registered trademarks of O'Reilly Media, Inc. *Hands-On Programming with R*, the picture of an orange-winged Amazon parrot, and related trade dress are trademarks of O'Reilly Media, Inc.

Many of the designations used by manufacturers and sellers to distinguish their products are claimed as trademarks. Where those designations appear in this book, and O'Reilly Media, Inc. was aware of a trademark claim, the designations have been printed in caps or initial caps.

While every precaution has been taken in the preparation of this book, the publisher and authors assume no responsibility for errors or omissions, or for damages resulting from the use of the information contained herein.

ISBN: 978-1-449-35901-0

[LSI]

Table of Contents

Foreword. vii
Preface. ix

Part I.   Project 1: Weighted Dice

1. The Very Basics. 3
    The R User Interface   3
    Objects   7
    Functions   12
    Sample with Replacement   14
    Writing Your Own Functions   16
        The Function Constructor   17
    Arguments   18
    Scripts   20
    Summary   22

2. Packages and Help Pages. 23
    Packages   23
        install.packages   24
        library   24
    Getting Help with Help Pages   29
        Parts of a Help Page   30
        Getting More Help   33
    Summary   33
    Project 1 Wrap-up   34

Part II. Project 2: Playing Cards

3. R Objects **37**
   Atomic Vectors   38
      Doubles   39
      Integers   40
      Characters   41
      Logicals   42
      Complex and Raw   42
   Attributes   43
      Names   44
      Dim   45
   Matrices   46
   Arrays   46
   Class   47
      Dates and Times   48
      Factors   49
   Coercion   51
   Lists   53
   Data Frames   55
   Loading Data   57
   Saving Data   61
   Summary   61

4. R Notation **65**
   Selecting Values   65
      Positive Integers   66
      Negative Integers   68
      Zero   69
      Blank Spaces   69
      Logical Values   69
      Names   70
   Deal a Card   70
   Shuffle the Deck   71
   Dollar Signs and Double Brackets   73
   Summary   76

5. Modifying Values **77**
   Changing Values in Place   77
   Logical Subsetting   80
      Logical Tests   80
      Boolean Operators   85

Missing Information	89
na.rm	90
is.na	90
Summary	91

6. Environments... 93
Environments	93
Working with Environments	95
The Active Environment	97
Scoping Rules	98
Assignment	99
Evaluation	99
Closures	108
Summary	112
Project 2 Wrap-up	113

Part III. Project 3: Slot Machine

7. Programs.. 117
Strategy	120
Sequential Steps	120
Parallel Cases	121
if Statements	122
else Statements	125
Lookup Tables	132
Code Comments	138
Summary	139

8. S3... 141
The S3 System	141
Attributes	142
Generic Functions	147
Methods	148
Method Dispatch	150
Classes	153
S3 and Debugging	154
S4 and R5	154
Summary	154

9. Loops.. 157
Expected Values	157

expand.grid	159
for Loops	165
while Loops	170
repeat Loops	171
Summary	171

10. Speed. 173

Vectorized Code	173
How to Write Vectorized Code	175
How to Write Fast for Loops in R	180
Vectorized Code in Practice	181
Loops Versus Vectorized Code	185
Summary	185
Project 3 Wrap-up	186

A. Installing R and RStudio. 189

B. R Packages. 193

C. Updating R and Its Packages. 197

D. Loading and Saving Data in R. 199

E. Debugging R Code. 213

Index. 223

Foreword

Learning to program is important if you're serious about understanding data. There's no argument that data science must be performed on a computer, but you have a choice between learning a graphical user interface (GUI) or a programming language. Both Garrett and I strongly believe that programming is a vital skill for everyone who works intensely with data. While convenient, a GUI is ultimately limiting, because it hampers three properties essential for good data analysis:

Reproducibility
   The ability to re-create a past analysis, which is crucial for good science.

Automation
   The ability to rapidly re-create an analysis when data changes (as it always does).

Communication
   Code is just text, so it is easy to communicate. When learning, this makes it easy to get help—whether it's with email, Google, Stack Overflow, or elsewhere.

Don't be afraid of programming! Anyone can learn to program with the right motivation, and this book is organized to keep you motivated. This is not a reference book; instead, it's structured around three hands-on challenges. Mastering these challenges will lead you through the basics of R programming and even into some intermediate topics, such as vectorized code, scoping, and S3 methods. Real challenges are a great way to learn, because you're not memorizing functions void of context; instead, you're learning functions as you need them to solve a real problem. You'll learn by doing, not by reading.

As you learn to program, you are going to get frustrated. You are learning a new language, and it will take time to become fluent. But frustration is not just natural, it's actually a positive sign that you should watch for. Frustration is your brain's way of being lazy; it's trying to get you to quit and go do something easy or fun. If you want to get physically fitter, you need to push your body even though it complains. If you want to get better at programming, you'll need to push your brain. Recognize when you get frustrated and see it as a good thing: you're now stretching yourself. Push yourself a little further every day, and you'll soon be a confident programmer.

Hands-On Programming with R is friendly, conversational, and active. It's the next-best thing to learning R programming from me or Garrett in person. I hope you enjoy reading it as much as I have.

—Hadley Wickham
Chief Scientist, RStudio

P.S. Garrett is too modest to mention it, but his lubridate package makes working with dates or times in R much less painful. Check it out!

Preface

This book will teach you how to program in R. You'll go from loading data to writing your own functions (which will outperform the functions of other R users). But this is not a typical introduction to R. I want to help you become a data scientist, as well as a computer scientist, so this book will focus on the programming skills that are most related to data science.

The chapters in the book are arranged according to three practical projects—given that they're fairly substantial projects, they span multiple chapters. I chose these projects for two reasons. First, they cover the breadth of the R language. You will learn how to load data, assemble and disassemble data objects, navigate R's environment system, write your own functions, and use all of R's programming tools, such as `if else` statements, for loops, S3 classes, R's package system, and R's debugging tools. The projects will also teach you how to write vectorized R code, a style of lightning-fast code that takes advantage of all of the things R does best.

But more importantly the projects will teach you how to solve the logistical problems of data science—and there are many logistical problems. When you work with data, you will need to store, retrieve, and manipulate large sets of values without introducing errors. As you work through the book, I will teach you not just how to program with R, but how to use the programming skills to support your work as a data scientist.

Not every programmer needs to be a data scientist, so not every programmer will find this book useful. You will find this book helpful if you're in one of the following categories:

1. You already use R as a statistical tool but would like to learn how to write your own functions and simulations with R.
2. You would like to teach yourself how to program, and you see the sense of learning a language related to data science.

One of the biggest surprises in this book is that I do not cover traditional applications of R, such as models and graphs; instead, I treat R purely as a programming language. Why this narrow focus? R is designed to be a tool that helps scientists analyze data. It has many excellent functions that make plots and fit models to data. As a result, many statisticians learn to use R as if it were a piece of software—they learn which functions do what they want, and they ignore the rest.

This is an understandable approach to learning R. Visualizing and modeling data are complicated skills that require a scientist's full attention. It takes expertise, judgement, and focus to extract reliable insights from a data set. I would not recommend that any data scientist distract herself with computer programming until she feels comfortable with the basic theory and practice of her craft. If you would like to learn the craft of data science, I recommend the forthcoming book *Data Science with R*, my companion volume to this book.

However, learning to program *should* be on every data scientist's to-do list. Knowing how to program will make you a more flexible analyst and augment your mastery of data science in every way. My favorite metaphor for describing this was introduced by Greg Snow on the R help mailing list in May 2006. Using the functions in R is like riding a bus. Writing programs in R is like driving a car.

> Busses are very easy to use, you just need to know which bus to get on, where to get on, and where to get off (and you need to pay your fare). Cars, on the other hand, require much more work: you need to have some type of map or directions (even if the map is in your head), you need to put gas in every now and then, you need to know the rules of the road (have some type of drivers license). The big advantage of the car is that it can take you a bunch of places that the bus does not go and it is quicker for some trips that would require transferring between busses.
>
> Using this analogy, programs like SPSS are busses, easy to use for the standard things, but very frustrating if you want to do something that is not already preprogrammed.
>
> R is a 4-wheel drive SUV (though environmentally friendly) with a bike on the back, a kayak on top, good walking and running shoes in the passenger seat, and mountain climbing and spelunking gear in the back.
>
> R can take you anywhere you want to go if you take time to learn how to use the equipment, but that is going to take longer than learning where the bus stops are in SPSS.
>
> — Greg Snow

Greg compares R to SPSS, but he assumes that you use the full powers of R; in other words, that you learn how to program in R. If you only use functions that preexist in R, you are using R like SPSS: it is a bus that can only take you to certain places.

This flexibility matters to data scientists. The exact details of a method or simulation will change from problem to problem. If you cannot build a method tailored to your situation, you may find yourself tempted to make unrealistic assumptions just so you can you use an ill-suited method that already exists.

This book will help you make the leap from bus to car. I have written it for beginning programmers. I do not talk about the theory of computer science—there are no discussions of big O( ) and little o( ) in these pages. Nor do I get into advanced details such as the workings of *lazy evaluation*. These things are interesting if you think of computer science at the theoretical level, but they are a distraction when you first learn to program.

Instead, I teach you how to program in R with three concrete examples. These examples are short, easy to understand, and cover everything you need to know.

I have taught this material many times in my job as Master Instructor at RStudio. As a teacher, I have found that students learn abstract concepts much faster when they are illustrated by concrete examples. The examples have a second advantage, as well: they provide immediate practice. Learning to program is like learning to speak another language—you progress faster when you practice. In fact, learning to program *is* learning to speak another language. You will get the best results if you follow along with the examples in the book and experiment whenever an idea strikes you.

> The book is a companion to *Data Science with R*. In that book, I explain how to use R to make plots, model data, and write reports. That book teaches these tasks as data-science skills, which require judgement and expertise—not as programming exercises, which they also are. This book will teach you how to program in R. It does not assume that you have mastered the data-science skills taught in volume 1 (nor that you ever intend to). However, this skill set amplifies that one. And if you master both, you will be a powerful, computer-augmented data scientist, fit to command a high salary and influence scientific dialogue.

Conventions Used in This Book

The following typographical conventions are used in this book:

Italic
: Indicates new terms, URLs, email addresses, filenames, and file extensions.

`Constant width`
: Used for program listings, as well as within paragraphs to refer to program elements such as variable or function names, databases, data types, environment variables, statements, and keywords.

`Constant width bold`
: Shows commands or other text that should be typed literally by the user.

`Constant width italic`
: Shows text that should be replaced with user-supplied values or by values determined by context.

This element signifies a tip or suggestion.

This element signifies a general note.

This element indicates a warning or caution.

Safari® Books Online

Safari Books Online is an on-demand digital library that delivers expert content in both book and video form from the world's leading authors in technology and business.

Technology professionals, software developers, web designers, and business and creative professionals use Safari Books Online as their primary resource for research, problem solving, learning, and certification training.

Safari Books Online offers a range of plans and pricing for enterprise, government, education, and individuals.

Members have access to thousands of books, training videos, and prepublication manuscripts in one fully searchable database from publishers like O'Reilly Media, Prentice Hall Professional, Addison-Wesley Professional, Microsoft Press, Sams, Que, Peachpit Press, Focal Press, Cisco Press, John Wiley & Sons, Syngress, Morgan Kaufmann, IBM Redbooks, Packt, Adobe Press, FT Press, Apress, Manning, New Riders, McGraw-Hill, Jones & Bartlett, Course Technology, and hundreds more. For more information about Safari Books Online, please visit us online.

How to Contact Us

Please address comments and questions concerning this book to the publisher:

O'Reilly Media, Inc.
1005 Gravenstein Highway North

Sebastopol, CA 95472
800-998-9938 (in the United States or Canada)
707-829-0515 (international or local)
707-829-0104 (fax)

We have a web page for this book, where we list errata, examples, and any additional information. You can access this page at *http://bit.ly/HandsOnR*

To comment or ask technical questions about this book, send email to *bookquestions@oreilly.com*.

For more information about our books, courses, conferences, and news, see our website at *http://www.oreilly.com*.

Find us on Facebook: *http://facebook.com/oreilly*

Follow us on Twitter: *http://twitter.com/oreillymedia*

Watch us on YouTube: *http://www.youtube.com/oreillymedia*

Acknowledgments

Many excellent people have helped me write this book, from my two editors, Courtney Nash and Julie Steele, to the rest of the O'Reilly team, who designed, proofread, and indexed the book. Also, Greg Snow generously let me quote him in this preface. I offer them all my heartfelt thanks.

I would also like to thank Hadley Wickham, who has shaped the way I think about and teach R. Many of the ideas in this book come from Statistics 405, a course that I helped Hadley teach when I was a PhD student at Rice University.

Further ideas came from the students and teachers of Introduction to Data Science with R, a workshop that I teach on behalf of RStudio. Thank you to all of you. I'd like to offer special thanks to my teaching assistants Josh Paulson, Winston Chang, Jaime Ramos, Jay Emerson, and Vivian Zhang.

Thank you also to JJ Allaire and the rest of my colleagues at RStudio who provide the RStudio IDE, a tool that makes it much easier to use, teach, and write about R.

Finally, I would like to thank my wife, Kristin, for her support and understanding while I wrote this book.

PART I
Project 1: Weighted Dice

Computers let you assemble, manipulate, and visualize data sets, all at speeds that would have wowed yesterday's scientists. In short, computers give you scientific superpowers! But you'll need to pick up some programming skills if you wish to fully utilize them.

As a data scientist who knows how to program, you will improve your ability to:

- Memorize (store) entire data sets
- Recall data values on demand
- Perform complex calculations with large amounts of data
- Do repetitive tasks without becoming careless or bored

Computers can do all of these things quickly and error free, which lets your mind do the things *it* excels at: making decisions and assigning meaning.

Sound exciting? Great! Let's begin.

When I was a college student, I sometimes daydreamed of going to Las Vegas. I thought that knowing statistics might help me win big. If that's what led *you* to data science, you better sit down; I have some bad news. Even a statistician will lose money in a casino over the long run. This is because the odds for each game are always stacked in the casino's favor. However, there is a loophole to this rule. You can make money—and reliably too. All you have to do is *be the casino*.

Believe it or not, R can help you do that. Over the course of the book, you will use R to build three virtual objects: a pair of dice that you can roll to generate random numbers, a deck of cards that you can shuffle and deal from, and a slot machine modeled after some real-life video lottery terminals. After that, you'll just need to add some video graphics and a bank account (and maybe get a few government licenses), and you'll be in business. I'll leave those details to you.

These projects are lighthearted, but they are also deep. As you complete them, you will become an expert at the skills you need to work with data as a data scientist. You will learn how to store data in your computer's memory, how to access data that is already there, and how to transform data values in memory when necessary. You will also learn how to write your own programs in R that you can use to analyze data and run simulations.

If simulating a slot machine (or dice, or cards) seems frivolous, think of it this way: playing a slot machine is a process. Once you can simulate it, you'll be able to simulate other processes, such as bootstrap sampling, Markov chain Monte Carlo, and other data-analysis procedures. Plus, these projects provide concrete examples for learning all the components of R programming: objects, data types, classes, notation, functions, environments, if trees, loops, and vectorization. This first project will make it easier to study these things by teaching you the basics of R.

Your first mission is simple: assemble R code that will simulate rolling a pair of dice, like at a craps table. Once you have done that, we'll weight the dice a bit in your favor, just to keep things interesting.

In this project, you will learn how to:

- Use the R and RStudio interfaces
- Run R commands
- Create R objects
- Write your own R functions and scripts
- Load and use R packages
- Generate random samples
- Create quick plots
- Get help when you need it

Don't worry if it seems like we cover a lot of ground fast. This project is designed to give you a concise overview of the R language. You will return to many of the concepts we meet here in projects 2 and 3, where you will examine the concepts in depth.

You'll need to have both R and RStudio installed on your computer before you can use them. Both are free and easy to download. See Appendix A for complete instructions. If you are ready to begin, open RStudio on your computer and read on.

CHAPTER 1
The Very Basics

This chapter provides a broad overview of the R language that will get you programming right away. In it, you will build a pair of virtual dice that you can use to generate random numbers. Don't worry if you've never programmed before; the chapter will teach you everything you need to know.

To simulate a pair of dice, you will have to distill each die into its essential features. You cannot place a physical object, like a die, into a computer (well, not without unscrewing some screws), but you can save *information* about the object in your computer's memory.

Which information should you save? In general, a die has six important pieces of information: when you roll a die, it can only result in one of six numbers: 1, 2, 3, 4, 5, and 6. You can capture the essential characteristics of a die by saving the numbers 1, 2, 3, 4, 5, and 6 as a group of values in your computer's memory.

Let's work on saving these numbers first and then consider a method for "rolling" our die.

The R User Interface

Before you can ask your computer to save some numbers, you'll need to know how to talk to it. That's where R and RStudio come in. RStudio gives you a way to talk to your computer. R gives you a language to speak in. To get started, open RStudio just as you would open any other application on your computer. When you do, a window should appear in your screen like the one shown in Figure 1-1.

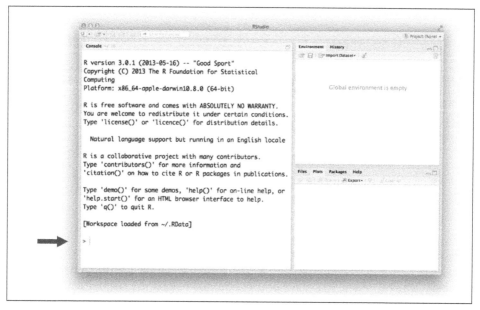

Figure 1-1. Your computer does your bidding when you type R commands at the prompt in the bottom line of the console pane. Don't forget to hit the Enter key. When you first open RStudio, the console appears in the pane on your left, but you can change this with File > Preferences in the menu bar.

 If you do not yet have R and RStudio intalled on your computer—or do not know what I am talking about—visit Appendix A. The appendix will give you an overview of the two free tools and tell you how to download them.

The RStudio interface is simple. You type R code into the bottom line of the RStudio console pane and then click Enter to run it. The code you type is called a *command*, because it will command your computer to do something for you. The line you type it into is called the *command line*.

When you type a command at the prompt and hit Enter, your computer executes the command and shows you the results. Then RStudio displays a fresh prompt for your next command. For example, if you type 1 + 1 and hit Enter, RStudio will display:

```
> 1 + 1
[1] 2
>
```

You'll notice that a [1] appears next to your result. R is just letting you know that this line begins with the first value in your result. Some commands return more than one

4 | Chapter 1: The Very Basics

value, and their results may fill up multiple lines. For example, the command 100:130 returns 31 values; it creates a sequence of integers from 100 to 130. Notice that new bracketed numbers appear at the start of the second and third lines of output. These numbers just mean that the second line begins with the 14th value in the result, and the third line begins with the 25th value. You can mostly ignore the numbers that appear in brackets:

```
> 100:130   ❶
 [1] 100 101 102 103 104 105 106 107 108 109 110 111 112
[14] 113 114 115 116 117 118 119 120 121 122 123 124 125
[25] 126 127 128 129 130
```

❶ The colon operator (:) returns every integer between two integers. It is an easy way to create a sequence of numbers.

Isn't R a language?

You may hear me speak of R in the third person. For example, I might say, "Tell R to do this" or "Tell R to do that", but of course R can't do anything; it is just a language. This way of speaking is shorthand for saying, "Tell your computer to do this by writing a command in the R language at the command line of your RStudio console." Your computer, and not R, does the actual work.

Is this shorthand confusing and slightly lazy to use? Yes. Do a lot of people use it? Everyone I know—probably because it is so convenient.

When do we compile?

In some languages, like C, Java, and FORTRAN, you have to compile your human-readable code into machine-readable code (often 1s and 0s) before you can run it. If you've programmed in such a language before, you may wonder whether you have to compile your R code before you can use it. The answer is no. R is a dynamic programming language, which means R automatically interprets your code as you run it.

If you type an incomplete command and press Enter, R will display a + prompt, which means it is waiting for you to type the rest of your command. Either finish the command or hit Escape to start over:

```
> 5 -
+
+ 1
[1] 4
```

If you type a command that R doesn't recognize, R will return an error message. If you ever see an error message, don't panic. R is just telling you that your computer couldn't understand or do what you asked it to do. You can then try a different command at the next prompt:

```
> 3 % 5
Error: unexpected input in "3 % 5"
>
```

Once you get the hang of the command line, you can easily do anything in R that you would do with a calculator. For example, you could do some basic arithmetic:

```
2 * 3
## 6

4 - 1
## 3

6 / (4 - 1)
## 2
```

Did you notice something different about this code? I've left out the >'s and [1]'s. This will make the code easier to copy and paste if you want to put it in your own console.

R treats the hashtag character, #, in a special way; R will not run anything that follows a hashtag on a line. This makes hashtags very useful for adding comments and annotations to your code. Humans will be able to read the comments, but your computer will pass over them. The hashtag is known as the *commenting symbol* in R.

For the remainder of the book, I'll use hashtags to display the output of R code. I'll use a single hashtag to add my own comments and a double hashtag, ##, to display the results of code. I'll avoid showing >s and [1]s unless I want you to look at them.

Cancelling commands

Some R commands may take a long time to run. You can cancel a command once it has begun by typing ctrl + c. Note that it may also take R a long time to cancel the command.

Exercise

That's the basic interface for executing R code in RStudio. Think you have it? If so, try doing these simple tasks. If you execute everything correctly, you should end up with the same number that you started with:

1. Choose any number and add 2 to it.
2. Multiply the result by 3.
3. Subtract 6 from the answer.

> 4. Divide what you get by 3.

Throughout the book, I'll put exercises in boxes, like the one just mentioned. I'll follow each exercise with a model answer, like the one that follows.

You could start with the number 10, and then do the preceding steps:

```
10 + 2
## 12

12 * 3
## 36

36 - 6
## 30

30 / 3
## 10
```

Now that you know how to use R, let's use it to make a virtual die. The : operator from a couple of pages ago gives you a nice way to create a group of numbers from one to six. The : operator returns its results as a *vector*, a one-dimensional set of numbers:

```
1:6
## 1 2 3 4 5 6
```

That's all there is to how a virtual die looks! But you are not done yet. Running 1:6 generated a vector of numbers for you to see, but it didn't save that vector anywhere in your computer's memory. What you are looking at is basically the footprints of six numbers that existed briefly and then melted back into your computer's RAM. If you want to use those numbers again, you'll have to ask your computer to save them somewhere. You can do that by creating an R *object*.

Objects

R lets you save data by storing it inside an R object. What's an object? Just a name that you can use to call up stored data. For example, you can save data into an object like *a* or *b*. Wherever R encounters the object, it will replace it with the data saved inside, like so:

```
a <- 1   ❶
a        ❷
## 1

a + 2    ❸
## 3
```

❶ To create an R object, choose a name and then use the less-than symbol, <, followed by a minus sign, -, to save data into it. This combination looks like an arrow, <-. R will make an object, give it your name, and store in it whatever follows the arrow.

❷ When you ask R what's in a, it tells you on the next line.

❸ You can use your object in new R commands, too. Since a previously stored the value of 1, you're now adding 1 to 2.

So, for another example, the following code would create an object named die that contains the numbers one through six. To see what is stored in an object, just type the object's name by itself:

```
die <- 1:6

die
## 1 2 3 4 5 6
```

When you create an object, the object will appear in the environment pane of RStudio, as shown in Figure 1-2. This pane will show you all of the objects you've created since opening RStudio.

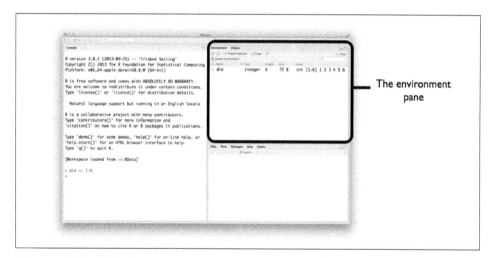

Figure 1-2. The RStudio environment pane keeps track of the R objects you create.

You can name an object in R almost anything you want, but there are a few rules. First, a name cannot start with a number. Second, a name cannot use some special symbols, like ^, !, $, @, +, -, /, or *:

Good names	Names that cause errors
a	1trial
b	$
FOO	^mean
my_var	2nd
.day	!bad

R also understands capitalization (or is case-sensitive), so name and Name will refer to different objects:

```
Name <- 1
name <- 0

Name + 1
## 2
```

Finally, R will overwrite any previous information stored in an object without asking you for permission. So, it is a good idea to *not* use names that are already taken:

```
my_number <- 1
my_number
## 1

my_number <- 999
my_number
## 999
```

You can see which object names you have already used with the function ls:

```
ls()
## "a"         "die"        "my_number" "name"      "Name"
```

You can also see which names you have used by examining RStudio's environment pane.

You now have a virtual die that is stored in your computer's memory. You can access it whenever you like by typing the word **die**. So what can you do with this die? Quite a lot. R will replace an object with its contents whenever the object's name appears in a command. So, for example, you can do all sorts of math with the die. Math isn't so helpful for rolling dice, but manipulating sets of numbers will be your stock and trade as a data scientist. So let's take a look at how to do that:

```
die - 1
## 0 1 2 3 4 5

die / 2
## 0.5 1.0 1.5 2.0 2.5 3.0

die * die
## 1  4  9 16 25 36
```

If you are a big fan of linear algebra (and who isn't?), you may notice that R does not always follow the rules of matrix multiplication. Instead, R uses *element-wise execution*. When you manipulate a set of numbers, R will apply the same operation to each element in the set. So for example, when you run **die - 1**, R subtracts one from each element of die.

When you use two or more vectors in an operation, R will line up the vectors and perform a sequence of individual operations. For example, when you run **die * die**, R lines up the two die vectors and then multiplies the first element of vector 1 by the first element of vector 2. It then multiplies the second element of vector 1 by the second element of vector 2, and so on, until every element has been multiplied. The result will be a new vector the same length as the first two, as shown in Figure 1-3.

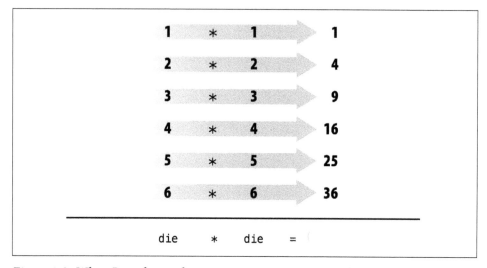

Figure 1-3. When R performs element-wise execution, it matches up vectors and then manipulates each pair of elements independently.

If you give R two vectors of unequal lengths, R will repeat the shorter vector until it is as long as the longer vector, and then do the math, as shown in Figure 1-4. This isn't a permanent change—the shorter vector will be its original size after R does the math. If the length of the short vector does not divide evenly into the length of the long vector, R will return a warning message. This behavior is known as *vector recycling*, and it helps R do element-wise operations:

    1:2
    ## 1 2

    1:4
    ## 1 2 3 4

```
die
## 1 2 3 4 5 6

die + 1:2
## 2 4 4 6 6 8

die + 1:4
## 2 4 6 8 6 8
Warning message:
In die + 1:4 :
  longer object length is not a multiple of shorter object length
```

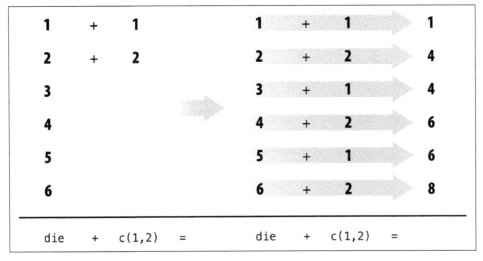

Figure 1-4. R will repeat a short vector to do element-wise operations with two vectors of uneven lengths.

Element-wise operations are a very useful feature in R because they manipulate groups of values in an orderly way. When you start working with data sets, element-wise operations will ensure that values from one observation or case are only paired with values from the same observation or case. Element-wise operations also make it easier to write your own programs and functions in R.

But don't think that R has given up on traditional matrix multiplication. You just have to ask for it when you want it. You can do inner multiplication with the %*% operator and outer multiplication with the %o% operator:

```
die %*% die
##      [,1]
## [1,]   91

die %o% die
##      [,1] [,2] [,3] [,4] [,5] [,6]
```

```
## [1,]    1    2    3    4    5    6
## [2,]    2    4    6    8   10   12
## [3,]    3    6    9   12   15   18
## [4,]    4    8   12   16   20   24
## [5,]    5   10   15   20   25   30
## [6,]    6   12   18   24   30   36
```

You can also do things like transpose a matrix with t and take its determinant with det.

Don't worry if you're not familiar with these operations. They are easy to look up, and you won't need them for this book.

Now that you can do math with your die object, let's look at how you could "roll" it. Rolling your die will require something more sophisticated than basic arithmetic; you'll need to randomly select one of the die's values. And for that, you will need a *function*.

Functions

R comes with many functions that you can use to do sophisticated tasks like random sampling. For example, you can round a number with the round function, or calculate its factorial with the factorial function. Using a function is pretty simple. Just write the name of the function and then the data you want the function to operate on in parentheses:

```
round(3.1415)
## 3

factorial(3)
## 6
```

The data that you pass into the function is called the function's *argument*. The argument can be raw data, an R object, or even the results of another R function. In this last case, R will work from the innermost function to the outermost, as in Figure 1-5:

```
mean(1:6)
## 3.5

mean(die)
## 3.5

round(mean(die))
## 4
```

Lucky for us, there is an R function that can help "roll" the die. You can simulate a roll of the die with R's sample function. sample takes *two* arguments: a vector named x and a number named size. sample will return size elements from the vector:

```
sample(x = 1:4, size = 2)
## 3 2
```

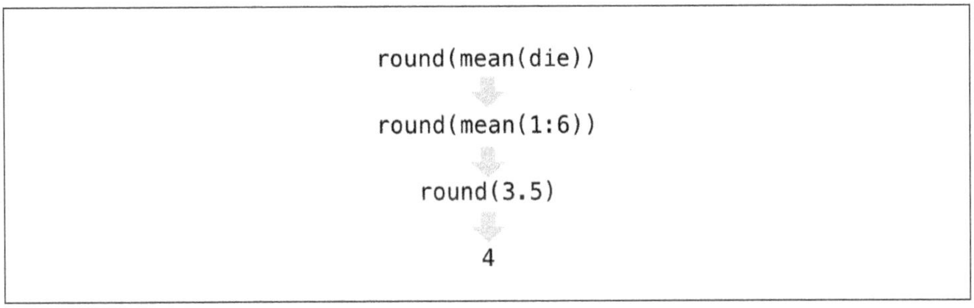

```
                    round(mean(die))

                    round(mean(1:6))

                       round(3.5)

                           4
```

Figure 1-5. When you link functions together, R will resolve them from the innermost operation to the outermost. Here R first looks up die, then calculates the mean of one through six, then rounds the mean.

To roll your die and get a number back, set x to die and sample one element from it. You'll get a new (maybe different) number each time you roll it:

```
sample(x = die, size = 1)
## 2

sample(x = die, size = 1)
## 1

sample(x = die, size = 1)
## 6
```

Many R functions take multiple arguments that help them do their job. You can give a function as many arguments as you like as long as you separate each argument with a comma.

You may have noticed that I set die and 1 equal to the names of the arguments in sample, x and size. Every argument in every R function has a name. You can specify which data should be assigned to which argument by setting a name equal to data, as in the preceding code. This becomes important as you begin to pass multiple arguments to the same function; names help you avoid passing the wrong data to the wrong argument. However, using names is optional. You will notice that R users do not often use the name of the first argument in a function. So you might see the previous code written as:

```
sample(die, size = 1)
## 2
```

Often, the name of the first argument is not very descriptive, and it is usually obvious what the first piece of data refers to anyways.

But how do you know which argument names to use? If you try to use a name that a function does not expect, you will likely get an error:

```
round(3.1415, corners = 2)
## Error in round(3.1415, corners = 2) : unused argument(s) (corners = 2)
```

If you're not sure which names to use with a function, you can look up the function's arguments with `args`. To do this, place the name of the function in the parentheses behind `args`. For example, you can see that the `round` function takes two arguments, one named x and one named `digits`:

```
args(round)
## function (x, digits = 0)
## NULL
```

Did you notice that `args` shows that the `digits` argument of `round` is already set to 0? Frequently, an R function will take optional arguments like `digits`. These arguments are considered optional because they come with a default value. You can pass a new value to an optional argument if you want, and R will use the default value if you do not. For example, `round` will round your number to 0 digits past the decimal point by default. To override the default, supply your own value for `digits`:

```
round(3.1415, digits = 2)
## 3.14
```

You should write out the names of each argument after the first one or two when you call a function with multiple arguments. Why? First, this will help you and others understand your code. It is usually obvious which argument your first input refers to (and sometimes the second input as well). However, you'd need a large memory to remember the third and fourth arguments of every R function. Second, and more importantly, writing out argument names prevents errors.

If you do not write out the names of your arguments, R will match your values to the arguments in your function by order. For example, in the following code, the first value, die, will be matched to the first argument of `sample`, which is named x. The next value, 1, will be matched to the next argument, `size`:

```
sample(die, 1)
## 2
```

As you provide more arguments, it becomes more likely that your order and R's order may not align. As a result, values may get passed to the wrong argument. Argument names prevent this. R will always match a value to its argument name, no matter where it appears in the order of arguments:

```
sample(size = 1, x = die)
## 2
```

Sample with Replacement

If you set `size = 2`, you can *almost* simulate a pair of dice. Before we run that code, think for a minute why that might be the case. `sample` will return two numbers, one for each die:

```
sample(die, size = 2)
## 3 4
```

I said this "almost" works because this method does something funny. If you use it many times, you'll notice that the second die never has the same value as the first die, which means you'll never roll something like a pair of threes or snake eyes. What is going on?

By default, `sample` builds a sample *without replacement*. To see what this means, imagine that `sample` places all of the values of `die` in a jar or urn. Then imagine that `sample` reaches into the jar and pulls out values one by one to build its sample. Once a value has been drawn from the jar, `sample` sets it aside. The value doesn't go back into the jar, so it cannot be drawn again. So if `sample` selects a six on its first draw, it will not be able to select a six on the second draw; six is no longer in the jar to be selected. Although `sample` creates its sample electronically, it follows this seemingly physical behavior.

One side effect of this behavior is that each draw depends on the draws that come before it. In the real world, however, when you roll a pair of dice, each die is independent of the other. If the first die comes up six, it does not prevent the second die from coming up six. In fact, it doesn't influence the second die in any way whatsoever. You can recreate this behavior in `sample` by adding the argument `replace = TRUE`:

```
sample(die, size = 2, replace = TRUE)
## 5 5
```

The argument `replace = TRUE` causes `sample` to sample *with replacement*. Our jar example provides a good way to understand the difference between sampling with replacement and without. When `sample` uses replacement, it draws a value from the jar and records the value. Then it puts the value back into the jar. In other words, `sample` *replaces* each value after each draw. As a result, `sample` may select the same value on the second draw. Each value has a chance of being selected each time. It is as if every draw were the first draw.

Sampling with replacement is an easy way to create *independent random samples*. Each value in your sample will be a sample of size one that is independent of the other values. This is the correct way to simulate a pair of dice:

```
sample(die, size = 2, replace = TRUE)
## 2 4
```

Congratulate yourself; you've just run your first simulation in R! You now have a method for simulating the result of rolling a pair of dice. If you want to add up the dice, you can feed your result straight into the `sum` function:

```
dice <- sample(die, size = 2, replace = TRUE)
dice
## 2 4

sum(dice)
## 6
```

What would happen if you call dice multiple times? Would R generate a new pair of dice values each time? Let's give it a try:

```
dice
## 2 4

dice
## 2 4

dice
## 2 4
```

Nope. Each time you call dice, R will show you the result of that one time you called sample and saved the output to dice. R won't rerun sample(die, 2, replace = TRUE) to create a new roll of the dice. This is a relief in a way. Once you save a set of results to an R object, those results do not change. Programming would be quite hard if the values of your objects changed each time you called them.

However, it *would* be convenient to have an object that can re-roll the dice whenever you call it. You can make such an object by writing your own R function.

Writing Your Own Functions

To recap, you already have working R code that simulates rolling a pair of dice:

```
die <- 1:6
dice <- sample(die, size = 2, replace = TRUE)
sum(dice)
```

You can retype this code into the console anytime you want to re-roll your dice. However, this is an awkward way to work with the code. It would be easier to use your code if you wrapped it into its own function, which is exactly what we'll do now. We're going to write a function named roll that you can use to roll your virtual dice. When you're finished, the function will work like this: each time you call roll(), R will return the sum of rolling two dice:

```
roll()
## 8

roll()
## 3

roll()
## 7
```

Functions may seem mysterious or fancy, but they are just another type of R object. Instead of containing data, they contain code. This code is stored in a special format that makes it easy to reuse the code in new situations. You can write your own functions by recreating this format.

The Function Constructor

Every function in R has three basic parts: a name, a body of code, and a set of arguments. To make your own function, you need to replicate these parts and store them in an R object, which you can do with the `function` function. To do this, call `function()` and follow it with a pair of braces, `{}`:

```
my_function <- function() {}
```

`function` will build a function out of whatever R code you place between the braces. For example, you can turn your dice code into a function by calling:

```
roll <- function() {
  die <- 1:6  ❶
  dice <- sample(die, size = 2, replace = TRUE)
  sum(dice)
}
```

❶ Notice that I indented each line of code between the braces. This makes the code easier for you and me to read but has no impact on how the code runs. R ignores spaces and line breaks and executes one complete expression at a time.

Just hit the Enter key between each line after the first brace, {. R will wait for you to type the last brace, }, before it responds.

Don't forget to save the output of `function` to an R object. This object will become your new function. To use it, write the object's name followed by an open and closed parenthesis:

```
roll()
## 9
```

You can think of the parentheses as the "trigger" that causes R to run the function. If you type in a function's name *without* the parentheses, R will show you the code that is stored inside the function. If you type in the name *with* the parentheses, R will run that code:

```
roll
## function() {
##   die <- 1:6
##   dice <- sample(die, size = 2, replace = TRUE)
##   sum(dice)
## }

roll()
## 6
```

The code that you place inside your function is known as the *body* of the function. When you run a function in R, R will execute all of the code in the body and then return the result of the last line of code. If the last line of code doesn't return a value, neither will your function, so you want to ensure that your final line of code returns a value. One way to check this is to think about what would happen if you ran the body of code line by line in the command line. Would R display a result after the last line, or would it not?

Here's some code that would display a result:

```
dice
1 + 1
sqrt(2)
```

And here's some code that would not:

```
dice <- sample(die, size = 2, replace = TRUE)
two <- 1 + 1
a <- sqrt(2)
```

Do you notice the pattern? These lines of code do not return a value to the command line; they save a value to an object.

Arguments

What if we removed one line of code from our function and changed the name die to bones, like this?

```
roll2 <- function() {
  dice <- sample(bones, size = 2, replace = TRUE)
  sum(dice)
}
```

Now I'll get an error when I run the function. The function needs the object bones to do its job, but there is no object named bones to be found:

```
roll2()
## Error in sample(bones, size = 2, replace = TRUE) :
##   object 'bones' not found
```

You can supply bones when you call roll2 if you make bones an argument of the function. To do this, put the name bones in the parentheses that follow function when you define roll2:

```
roll2 <- function(bones) {
  dice <- sample(bones, size = 2, replace = TRUE)
  sum(dice)
}
```

Now roll2 will work as long as you supply bones when you call the function. You can take advantage of this to roll different types of dice each time you call roll2. Dungeons and Dragons, here we come!

Remember, we're rolling pairs of dice:

```
roll2(bones = 1:4)
## 3

roll2(bones = 1:6)
## 10

roll2(1:20)
## 31
```

Notice that `roll2` will still give an error if you do not supply a value for the `bones` argument when you call `roll2`:

```
roll2()
## Error in sample(bones, size = 2, replace = TRUE) :
##   argument "bones" is missing, with no default
```

You can prevent this error by giving the `bones` argument a default value. To do this, set `bones` equal to a value when you define `roll2`:

```
roll2 <- function(bones = 1:6) {
  dice <- sample(bones, size = 2, replace = TRUE)
  sum(dice)
}
```

Now you can supply a new value for `bones` if you like, and `roll2` will use the default if you do not:

```
roll2()
## 9
```

You can give your functions as many arguments as you like. Just list their names, separated by commas, in the parentheses that follow `function`. When the function is run, R will replace each argument name in the function body with the value that the user supplies for the argument. If the user does not supply a value, R will replace the argument name with the argument's default value (if you defined one).

To summarize, `function` helps you construct your own R functions. You create a body of code for your function to run by writing code between the braces that follow `function`. You create arguments for your function to use by supplying their names in the parentheses that follow `function`. Finally, you give your function a name by saving its output to an R object, as shown in Figure 1-6.

Once you've created your function, R will treat it like every other function in R. Think about how useful this is. Have you ever tried to create a new Excel option and add it to Microsoft's menu bar? Or a new slide animation and add it to Powerpoint's options? When you work with a programming language, you can do these types of things. As you learn to program in R, you will be able to create new, customized, reproducible tools for yourself whenever you like. Part III will teach you much more about writing functions in R.

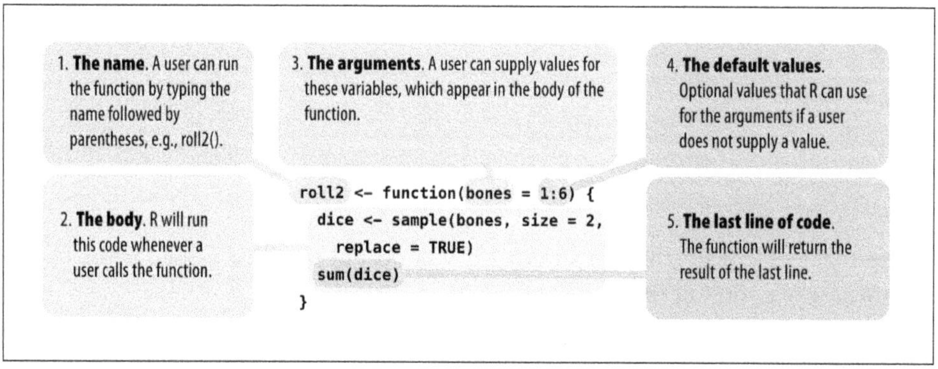

```
1. The name. A user can run        3. The arguments. A user can supply values for    4. The default values.
   the function by typing the         these variables, which appear in the body of the   Optional values that R can use
   name followed by                   function.                                           for the arguments if a user
   parentheses, e.g., roll2().                                                            does not supply a value.

                                   roll2 <- function(bones = 1:6) {
2. The body. R will run              dice <- sample(bones, size = 2,    5. The last line of code.
   this code whenever a                replace = TRUE)                     The function will return the
   user calls the function.          sum(dice)                             result of the last line.
                                   }
```

Figure 1-6. Every function in R has the same parts, and you can use function to create these parts.

Scripts

What if you want to edit `roll2` again? You could go back and retype each line of code in `roll2`, but it would be so much easier if you had a draft of the code to start from. You can create a draft of your code as you go by using an R *script*. An R script is just a plain text file that you save R code in. You can open an R script in RStudio by going to File > New File > R script in the menu bar. RStudio will then open a fresh script above your console pane, as shown in Figure 1-7.

I strongly encourage you to write and edit all of your R code in a script before you run it in the console. Why? This habit creates a reproducible record of your work. When you're finished for the day, you can save your script and then use it to rerun your entire analysis the next day. Scripts are also very handy for editing and proofreading your code, and they make a nice copy of your work to share with others. To save a script, click the scripts pane, and then go to File > Save As in the menu bar.

RStudio comes with many built-in features that make it easy to work with scripts. First, you can automatically execute a line of code in a script by clicking the Run button, as shown in Figure 1-8.

R will run whichever line of code your cursor is on. If you have a whole section highlighted, R will run the highlighted code. Alternatively, you can run the entire script by clicking the Source button. Don't like clicking buttons? You can use Control + Return as a shortcut for the Run button. On Macs, that would be Command + Return.

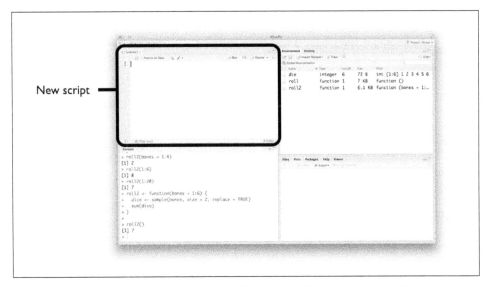

Figure 1-7. When you open an R Script (File > New File > R Script in the menu bar), RStudio creates a fourth pane above the console where you can write and edit your code.

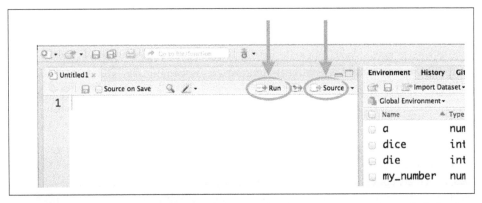

Figure 1-8. You can run a highlighted portion of code in your script if you click the Run button at the top of the scripts pane. You can run the entire script by clicking the Source button.

If you're not convinced about scripts, you soon will be. It becomes a pain to write multi-line code in the console's single-line command line. Let's avoid that headache and open your first script now before we move to the next chapter.

Scripts | 21

Extract function
RStudio comes with a tool that can help you build functions. To use it, highlight the lines of code in your R script that you want to turn into a function. Then click `Code > Extract Function` in the menu bar. RStudio will ask you for a function name to use and then wrap you code in a `function` call. It will scan the code for undefined variables and use these as arguments.

You may want to double-check RStudio's work. It assumes that your code is correct, so if it does something surprising, you may have a problem in your code.

Summary

You've covered a lot of ground already. You now have a virtual die stored in your computer's memory, as well as your own R function that rolls a pair of dice. You've also begun speaking the R language.

As you've seen, R is a language that you can use to talk to your computer. You write commands in R and run them at the command line for your computer to read. Your computer will sometimes talk back—for example, when you commit an error—but it usually just does what you ask and then displays the result.

The two most important components of the R language are objects, which store data, and functions, which manipulate data. R also uses a host of operators like +, -, *, /, and <- to do basic tasks. As a data scientist, you will use R objects to store data in your computer's memory, and you will use functions to automate tasks and do complicated calculations. We will examine objects in more depth later in Part II and dig further into functions in Part III. The vocabulary you have developed here will make each of those projects easier to understand. However, we're not done with your dice yet.

In Chapter 2, you'll run some simulations on your dice and build your first graphs in R. You'll also look at two of the most useful components of the R language: R *packages*, which are collections of functions writted by R's talented community of developers, and R documentation, which is a collection of help pages built into R that explains every function and data set in the language.

CHAPTER 2
Packages and Help Pages

You now have a function that simulates rolling a pair of dice. Let's make things a little more interesting by weighting the dice in your favor. The house always wins, right? Let's make the dice roll high numbers slightly more often than it rolls low numbers.

Before we weight the dice, we should make sure that they are fair to begin with. Two tools will help you do this: *repetition* and *visualization*. By coincidence, these tools are also two of the most useful superpowers in the world of data science.

We will repeat our dice rolls with a function called `replicate`, and we will visualize our rolls with a function called `qplot`. `qplot` does not come with R when you download it; `qplot` comes in a standalone R package. Many of the most useful R tools come in R packages, so let's take a moment to look at what R packages are and how you can use them.

Packages

You're not the only person writing your own functions with R. Many professors, programmers, and statisticians use R to design tools that can help people analyze data. They then make these tools free for anyone to use. To use these tools, you just have to download them. They come as preassembled collections of functions and objects called packages. Appendix B contains detailed instructions for downloading and updating R packages, but we'll look at the basics here.

We're going to use the `qplot` function to make some quick plots. `qplot` comes in the *ggplot2* package, a popular package for making graphs. Before you can use `qplot`, or anything else in the ggplot2 package, you need to download and install it.

install.packages

Each R package is hosted at *http://cran.r-project.org*, the same website that hosts R. However, you don't need to visit the website to download an R package; you can download packages straight from R's command line. Here's how:

1. Open RStudio.
2. Make sure you are connected to the Internet.
3. Run **install.packages("ggplot2")** at the command line.

That's it. R will have your computer visit the website, download ggplot2, and install the package in your hard drive right where R wants to find it. You now have the ggplot2 package. If you would like to install another package, replace ggplot2 with your package name in the code.

library

Installing a package doesn't place its functions at your fingertips just yet: it simply places them in your hard drive. To use an R package, you next have to load it in your R session with the command **library("ggplot2")**. If you would like to load a different package, replace ggplot2 with your package name in the code.

To see what this does, try an experiment. First, ask R to show you the `qplot` function. R won't be able to find `qplot` because `qplot` lives in the ggplot2 package, which you haven't loaded:

```
qplot
## Error: object 'qplot' not found
```

Now load the ggplot2 package:

```
library("ggplot2")
```

If you installed the package with `install.packages` as instructed, everything should go fine. Don't worry if you don't see any results or messages. No news is fine news when loading a package. Don't worry if you do see a message either; ggplot2 sometimes displays helpful start up messages. As long as you do not see anything that says "Error," you are doing fine.

Now if you ask to see `qplot`, R will show you quite a bit of code (`qplot` is a long function):

```
qplot
## (quite a bit of code)
```

Appendix B contains many more details about acquiring and using packages. I recommend that you read it if you are unfamiliar with R's package system. The main thing to remember is that you only need to install a package once, but you need to load it with library each time you wish to use it in a new R session. R will unload all of its packages each time you close RStudio.

Now that you've loaded qplot, let's take it for a spin. qplot makes "quick plots." If you give qplot two vectors of equal lengths, qplot will draw a scatterplot for you. qplot will use the first vector as a set of x values and the second vector as a set of y values. Look for the plot to appear in the Plots tab of the bottom-right pane in your RStudio window.

The following code will make the plot that appears in Figure 2-1. Until now, we've been creating sequences of numbers with the : operator; but you can also create vectors of numbers with the c function. Give c all of the numbers that you want to appear in the vector, separated by a comma. c stands for *concatenate*, but you can think of it as "collect" or "combine":

```
x <- c(-1, -0.8, -0.6, -0.4, -0.2, 0, 0.2, 0.4, 0.6, 0.8, 1)
x
## -1.0 -0.8 -0.6 -0.4 -0.2  0.0  0.2  0.4  0.6  0.8  1.0

y <- x^3
y
## -1.000 -0.512 -0.216 -0.064 -0.008  0.000  0.008
##  0.064  0.216  0.512  1.000

qplot(x, y)
```

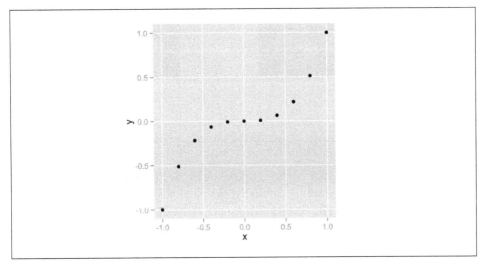

Figure 2-1. qplot makes a scatterplot when you give it two vectors.

You don't need to name your vectors x and y. I just did that to make the example clear. As you can see in Figure 2-1, a scatterplot is a set of points, each plotted according to its x and y values. Together, the vectors x and y describe a set of 10 points. How did R match up the values in x and y to make these points? With element-wise execution, as we saw in Figure 1-3.

Scatterplots are useful for visualizing the relationship between two variables. However, we're going to use a different type of graph, a *histogram*. A histogram visualizes the distribution of a single variable; it displays how many data points appear at each value of x.

Let's take a look at a histogram to see if this makes sense. qplot will make a histogram whenever you give it only *one* vector to plot. The following code makes the left-hand plot in Figure 2-2 (we'll worry about the right-hand plot in just second). To make sure our graphs look the same, use the extra argument **binwidth = 1**:

```
x <- c(1, 2, 2, 2, 3, 3)
qplot(x, binwidth = 1)
```

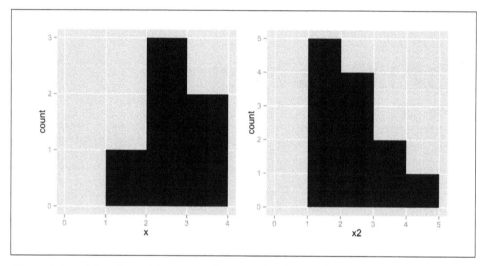

Figure 2-2. qplot makes a histogram when you give it a single vector.

This plot shows that our vector contains one value in the interval [1, 2) by placing a bar of height 1 above that interval. Similarly, the plot shows that the vector contains three values in the interval [2, 3) by placing a bar of height 3 in that interval. It shows that the vector contains two values in the interval [3, 4) by placing a bar of height 2 in that interval. In these intervals, the hard bracket, [, means that the first number is included in the interval. The parenthesis,), means that the last number is *not* included.

Let's try another histogram. This code makes the right-hand plot in Figure 2-2. Notice that there are five points with a value of 1 in x2. The histogram displays this by plotting a bar of height 5 above the interval x2 = [1, 2):

```
x2 <- c(1, 1, 1, 1, 1, 2, 2, 2, 2, 3, 3, 4)
qplot(x2, binwidth = 1)
```

> **Exercise**
>
> Let x3 be the following vector:
>
> ```
> x3 <- c(0, 1, 1, 2, 2, 2, 3, 3, 4)
> ```
>
> Imagine what a histogram of x3 would look like. Assume that the histogram has a bin width of 1. How many bars will the histogram have? Where will they appear? How high will each be?
>
> When you are done, plot a histogram of x3 with binwidth = 1, and see if you are right.

You can make a histogram of x3 with qplot(x3, binwidth = 1). The histogram will look like a symmetric pyramid. The middle bar will have a height of 3 and will appear above [2, 3), but be sure to try it and see for yourself.

You can use a histogram to display visually how common different values of x are. Numbers covered by a tall bar are more common than numbers covered by a short bar.

How can you use a histogram to check the accuracy of your dice?

Well, if you roll your dice many times and keep track of the results, you would expect some numbers to occur more than others. This is because there are more ways to get some numbers by adding two dice together than to get other numbers, as shown in Figure 2-3.

If you roll your dice many times and plot the results with qplot, the histogram will show you how often each sum appeared. The sums that occurred most often will have the highest bars. The histogram should look like the pattern in Figure 2-3 if the dice are fairly weighted.

This is where replicate comes in. replicate provides an easy way to repeat an R command many times. To use it, first give replicate the number of times you wish to repeat an R command, and then give it the command you wish to repeat. replicate will run the command multiple times and store the results as a vector:

```
replicate(3, 1 + 1)
## 2 2 2

replicate(10, roll())
## 3  7  5  3  6  2  3  8 11  7
```

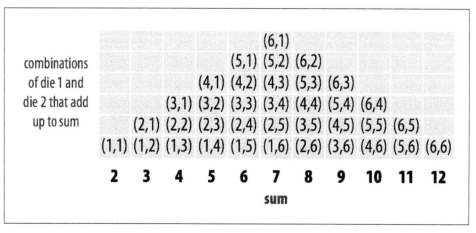

Figure 2-3. Each individual dice combination should occur with the same frequency. As a result, some sums will occur more often than others. With fair dice, each sum should appear in proportion to the number of combinations that make it.

A histogram of your first 10 rolls probably won't look like the pattern shown in Figure 2-3. Why not? There is too much randomness involved. Remember that we use dice in real life because they are effective random number generators. Patterns of long run frequencies will only appear *over the long run*. So let's simulate 10,000 dice rolls and plot the results. Don't worry; qplot and replicate can handle it. The results appear in Figure 2-4:

```
rolls <- replicate(10000, roll())
qplot(rolls, binwidth = 1)
```

The results suggest that the dice are fair. Over the long run, each number occurs in proportion to the number of combinations that generate it.

Now how can you bias these results? The previous pattern occurs because each underlying combination of dice (e.g., (3,4)) occurs with the same frequency. If you could increase the probability that a 6 is rolled on either die, then any combination with a six in it will occur more often than any combination without a six in it. The combination (6, 6) would occur most of all. This won't make the dice add up to 12 more often than they add up to seven, but it will skew the results toward the higher numbers.

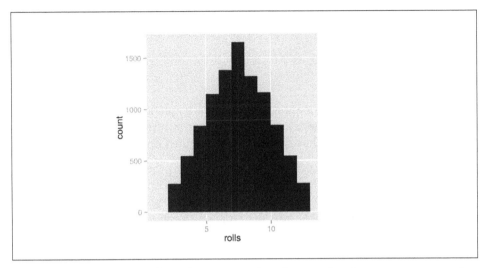

Figure 2-4. The behavior of our dice suggests that they are fair. Seven occurs more often than any other number, and frequencies diminish in proportion to the number of die combinations that create each number.

To put it another way, the probability of rolling any single number on a fair die is 1/6. I'd like you to change the probability to 1/8 for each number below six, and then increase the probability of rolling a six to 3/8:

Number	Fair probability	Weighted probability
1	1/6	1/8
2	1/6	1/8
3	1/6	1/8
4	1/6	1/8
5	1/6	1/8
6	1/6	3/8

You can change the probabilities by adding a new argument to the `sample` function. I'm not going to tell you what the argument is; instead I'll point you to the help page for the `sample` function. What's that? R functions come with help pages? Yes they do, so let's learn how to read one.

Getting Help with Help Pages

There are over 1,000 functions at the core of R, and new R functions are created all of the time. This can be a lot of material to memorize and learn! Luckily, each R function comes with its own help page, which you can access by typing the function's name after a question mark. For example, each of these commands will open a help page. Look for the pages to appear in the Help tab of RStudio's bottom-right pane:

```
?sqrt
?log10
?sample
```

Help pages contain useful information about what each function does. These help pages also serve as code documentation, so reading them can be bittersweet. They often seem to be written for people who already understand the function and do not need help.

Don't let this bother you—you can gain a lot from a help page by scanning it for information that makes sense and glossing over the rest. This technique will inevitably bring you to the most helpful part of each help page: the bottom. Here, almost every help page includes some example code that puts the function in action. Running this code is a great way to learn by example.

 If a function comes in an R package, R won't be able to find its help page unless the package is loaded.

Parts of a Help Page

Each help page is divided into sections. Which sections appear can vary from help page to help page, but you can usually expect to find these useful topics:

Description
    A short summary of what the function does.

Usage
    An example of how you would type the function. Each argument of the function will appear in the order R expects you to supply it (if you don't use argument names).

Arguments
    A list of each argument the function takes, what type of information R expects you to supply for the argument, and what the function will do with the information.

Details
    A more in-depth description of the function and how it operates. The details section also gives the function author a chance to alert you to anything you might want to know when using the function.

Value
    A description of what the function returns when you run it.

See Also
> A short list of related R functions.

Examples
> Example code that uses the function and is guaranteed to work. The examples section of a help page usually demonstrates a couple different ways to use a function. This helps give you an idea of what the function is capable of.

If you'd like to look up the help page for a function but have forgotten the function's name, you can search by keyword. To do this, type two question marks followed by a keyword in R's command line. R will pull up a list of links to help pages related to the keyword. You can think of this as the help page for the help page:

```
??log
```

Let's take a stroll through `sample`'s help page. Remember: we're searching for anything that could help you change the probabilities involved in the sampling process. I'm not going to reproduce the whole help page here (just the juiciest parts), so you should follow along on your computer.

First, open the help page. It will appear in the same pane in RStudio as your plots did (but in the Help tab, not the Plots tab):

```
?sample
```

What do you see? Starting from the top:

```
Random Samples and Permutations

Description
    sample takes a sample of the specified size from the elements of x using
either with or without replacement.
```

So far, so good. You knew all of that. The next section, Usage, has a possible clue. It mentions an argument called `prob`:

```
Usage
    sample(x, size, replace = FALSE, prob = NULL)
```

If you scroll down to the arguments section, the description of `prob` sounds *very* promising:

```
A vector of probability weights for obtaining the elements of the vector being
sampled.
```

The Details section confirms our suspicions. In this case, it also tells you how to proceed:

```
The optional prob argument can be used to give a vector of weights for obtaining
the elements of the vector being sampled. They need not sum to one, but they
should be nonnegative and not all zero.
```

Although the help page does not say it here, these weights will be matched up to the elements being sampled in element-wise fashion. The first weight will describe the first element, the second weight the second element, and so on. This is common practice in R.

Reading on:

If replace is true, Walker's alias method (Ripley, 1987) is used...

Okay, that looks like time to start skimming. We should have enough info now to figure out how to weight our dice.

Exercise

Rewrite the `roll` function to roll a pair of weighted dice:

```
roll <- function() {
  die <- 1:6
  dice <- sample(die, size = 2, replace = TRUE)
  sum(dice)
}
```

You will need to add a `prob` argument to the `sample` function inside of `roll`. This argument should tell `sample` to sample the numbers one through five with probability 1/8 and the number 6 with probability 3/8.

When you are finished, read on for a model answer.

To weight your dice, you need to add a `prob` argument with a vector of weights to `sample`, like this:

```
roll <- function() {
  die <- 1:6
  dice <- sample(die, size = 2, replace = TRUE,
    prob = c(1/8, 1/8, 1/8, 1/8, 1/8, 3/8))
  sum(dice)
}
```

This will cause `roll` to pick 1 through 5 with probability 1/8 and 6 with probability 3/8.

Overwrite your previous version of `roll` with the new function (by running the previous code snippet in your command line). Then visualize the new long-term behavior of your dice. I've put the results in Figure 2-5 next to our original results:

```
rolls <- replicate(10000, roll())
qplot(rolls, binwidth = 1)
```

This confirms that we've effectively weighted the dice. High numbers occur much more often than low numbers. The remarkable thing is that this behavior will only be apparent when you examine long-term frequencies. On any single roll, the dice will appear to behave randomly. This is great news if you play Settlers of Catan (just tell your friends you lost the dice), but it should be disturbing if you analyze data, because it means that bias can easily occur without anyone noticing it in the short run.

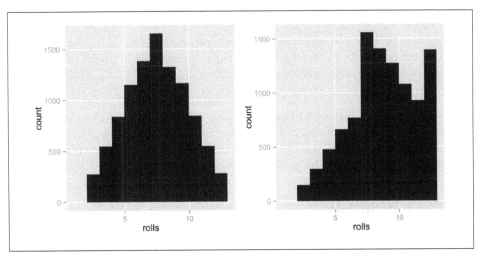

Figure 2-5. The dice are now clearly biased towards high numbers, since high sums occur much more often than low sums.

Getting More Help

R also comes with a super active community of users that you can turn to for help on the R-help mailing list (*http://bit.ly/r-help*). You can email the list with questions, but there's a great chance that your question has already been answered. Find out by searching the archives (*http://bit.ly/R_archives*).

Even better than the R-help list is Stack Overflow (*http://stackoverflow.com*), a website that allows programmers to answer questions and users to rank answers based on helpfulness. Personally, I find the Stack Overflow format to be more user-friendly than the R-help email list (and the respondents to be more human friendly). You can submit your own question or search through Stack Overflow's previously answered questions related to R. There are over 30,000.

For both the R help list and Stack Overflow, you're more likely to get a useful answer if you provide a reproducible example with your question. This means pasting in a short snippet of code that users can run to arrive at the bug or question you have in mind.

Summary

R's packages and help pages can make you a more productive programmer. You saw in Chapter 1 that R gives you the power to write your own functions that do specific things, but often the function that you want to write will already exist in an R package.

Professors, programmers, and scientists have developed over 5,000 packages for you to use, which can save you valuable programming time. To use a package, you need to install it to your computer once with `install.packages`, and then load it into each new R session with `library`.

R's help pages will help you master the functions that appear in R and its packages. Each function and data set in R has its own help page. Although help pages often contain advanced content, they also contain valuable clues and examples that can help you learn how to use a function.

You have now seen enough of R to learn by doing, which is the best way to learn R. You can make your own R commands, run them, and get help when you need to understand something that I have not explained. I encourage you to experiment with your own ideas in R as you read through the next two projects.

Project 1 Wrap-up

You've done more in this project than enable fraud and gambling; you've also learned how to speak to your computer in the language of R. R is a language like English, Spanish, or German, except R helps you talk to computers, not humans.

You've met the nouns of the R language, objects. And hopefully you guessed that functions are the verbs (I suppose function arguments would be the adverbs). When you combine functions and objects, you express a complete thought. By stringing thoughts together in a logical sequence, you can build eloquent, even artistic statements. In that respect, R is not that different than any other language.

R shares another characteristic of human languages: you won't feel very comfortable speaking R until you build up a vocabulary of R commands to use. Fortunately, you don't have to be bashful. Your computer will be the only one to "hear" you speak R. Your computer is not very forgiving, but it also doesn't judge. Not that you need to worry; you'll broaden your R vocabulary tremendously between here and the end of the book.

Now that you can use R, it is time to become an expert at using R to do data science. The foundation of data science is the ability to store large amounts of data and recall values on demand. From this, all else follows—manipulating data, visualizing data, modeling data, and more. However, you cannot easily store a data set in your mind by memorizing it. Nor can you easily store a data set on paper by writing it down. The only efficient way to store large amounts of data is with a computer. In fact, computers are so efficient that their development over the last three decades has completely changed the type of data we can accumulate and the methods we can use to analyze it. In short, computer data storage has driven the revolution in science that we call data science.

Part II will make you part of this revolution by teaching you how to use R to store data sets in your computer's memory and how to retrieve and manipulate data once it's there.

PART II
Project 2: Playing Cards

This project—which spans the next four chapters—will teach you how to store, retrieve, and change data values in your computer's memory. These skills will help you save and manage data without accumulating errors. In the project, you'll design a deck of playing cards that you can shuffle and deal from. Best of all, the deck will remember which cards have been dealt—just like a real deck. You can use the deck to play card games, tell fortunes, and test card-counting strategies.

Along the way, you will learn how to:

- Save new types of data, like character strings and logical values
- Save a data set as a vector, matrix, array, list, or data frame
- Load and save your own data sets with R
- Extract individual values from a data set
- Change individual values within a data set
- Write logical tests
- Use R's missing-value symbol, NA

To keep the project simple, I've divided it into four tasks. Each task will teach you a new skill for managing data with R:

Task 1: build the deck
    In Chapter 3, you will design and build a virtual deck of playing cards. This will be a complete data set, just like the ones you will use as a data scientist. You'll need to know how to use R's data types and data structures to make this work.

Task 2: write functions that deal and shuffle
   Next, in Chapter 4, you will write two functions to use with the deck. One function will deal cards from the deck, and the other will reshuffle the deck. To write these functions, you'll need to know how to extract values from a data set with R.

Task 3: change the point system to suit your game
   In Chapter 5, you will use R's notation system to change the point values of your cards to match the card games you may wish to play, like war, hearts, or blackjack. This will help you change values in place in existing data sets.

Task 4: manage the state of the deck
   Finally, in Chapter 6, you will make sure that your deck remembers which cards it has dealt. This is an advanced task, and it will introduce R's environment system and scoping rules. To do it successfully, you will need to learn the minute details of how R looks up and uses the data that you have stored in your computer.

CHAPTER 3
R Objects

In this chapter, you'll use R to assemble a deck of 52 playing cards.

You'll start by building simple R objects that represent playing cards and then work your way up to a full-blown table of data. In short, you'll build the equivalent of an Excel spreadsheet from scratch. When you are finished, your deck of cards will look something like this:

```
    face   suit value
    king spades    13
   queen spades    12
    jack spades    11
     ten spades    10
    nine spades     9
   eight spades     8
   ...
```

Do you need to build a data set from scratch to use it in R? Not at all. You can load most data sets into R with one simple step, see "Loading Data" on page 57. But this exercise will teach you how R stores data, and how you can assemble—or disassemble—your own data sets. You will also learn about the various types of objects available for you to use in R (not all R objects are the same!). Consider this exercise a rite of passage; by doing it, you will become an expert on storing data in R.

We'll start with the very basics. The most simple type of object in R is an *atomic vector*. Atomic vectors are not nuclear powered, but they are very simple and they do show up everywhere. If you look closely enough, you'll see that most structures in R are built from atomic vectors.

Atomic Vectors

An atomic vector is just a simple vector of data. In fact, you've already made an atomic vector, your die object from Part I. You can make an atomic vector by grouping some values of data together with c:

```
die <- c(1, 2, 3, 4, 5, 6)
die
## 1 2 3 4 5 6

is.vector(die)   ❶
##   TRUE
```

❶ is.vector tests whether an object is an atomic vector. It returns TRUE if the object is an atomic vector and FALSE otherwise.

You can also make an atomic vector with just one value. R saves single values as an atomic vector of length 1:

```
five <- 5
five
## 5

is.vector(five)
##   TRUE

length(five)
## 1
length(die)
## 6
```

length
length returns the length of an atomic vector.

Each atomic vector stores its values as a one-dimensional vector, and each atomic vector can only store one type of data. You can save different types of data in R by using different types of atomic vectors. Altogether, R recognizes six basic types of atomic vectors: *doubles, integers, characters, logicals, complex,* and *raw*.

To create your card deck, you will need to use different types of atomic vectors to save different types of information (text and numbers). You can do this by using some simple conventions when you enter your data. For example, you can create an integer vector by including a capital L with your input. You can create a character vector by surrounding your input in quotation marks:

```
int <- 1L
text <- "ace"
```

Each type of atomic vector has its own convention (described below). R will recognize the convention and use it to create an atomic vector of the appropriate type. If you'd like to make atomic vectors that have more than one element in them, you can combine an element with the c function from Chapter 2. Use the same convention with each element:

```
int <- c(1L, 5L)
text <- c("ace", "hearts")
```

You may wonder why R uses multiple types of vectors. Vector types help R behave as you would expect. For example, R will do math with atomic vectors that contain numbers, but not with atomic vectors that contain character strings:

```
sum(int)
## 6

sum(text)
## Error in sum(text) : invalid 'type' (character) of argument
```

But we're getting ahead of ourselves! Get ready to say hello to the six types of atomic vectors in R.

Doubles

A double vector stores regular numbers. The numbers can be positive or negative, large or small, and have digits to the right of the decimal place or not. In general, R will save any number that you type in R as a double. So, for example, the die you made in Part I was a double object:

```
die <- c(1, 2, 3, 4, 5, 6)
die
## 1 2 3 4 5 6
```

You'll usually know what type of object you are working with in R (it will be obvious), but you can also ask R what type of object an object is with typeof. For example:

```
typeof(die)
## "double"
```

Some R functions refer to doubles as "numerics," and I will often do the same. Double is a computer science term. It refers to the specific number of bytes your computer uses to store a number, but I find "numeric" to be much more intuitive when doing data science.

Integers

Integer vectors store integers, numbers that can be written without a decimal component. As a data scientist, you won't use the integer type very often because you can save integers as a double object.

You can specifically create an integer in R by typing a number followed by an uppercase L. For example:

```
int <- c(-1L, 2L, 4L)
int
## -1  2  4

typeof(int)
## "integer"
```

Note that R won't save a number as an integer unless you include the L. Integer numbers without the L will be saved as doubles. The only difference between 4 and 4L is how R saves the number in your computer's memory. Integers are defined more precisely in your computer's memory than doubles (unless the integer is *very* large).

Why would you save your data as an integer instead of a double? Sometimes a difference in precision can have surprising effects. Your computer allocates 64 bits of memory to store each double in an R program. This allows a lot of precision, but some numbers cannot be expressed exactly in 64 bits, the equivalent of a sequence of 64 ones and zeroes. For example, the number π contains an endless sequences of digits to the right of the decimal place. Your computer must round π to something close to, but not exactly equal to π to store π in its memory. Many decimal numbers share a similar fate.

As a result, each double is accurate to about 16 significant digits. This introduces a little bit of error. In most cases, this rounding error will go unnoticed. However, in some situations, the rounding error can cause surprising results. For example, you may expect the result of the expression below to be zero, but it is not:

```
sqrt(2)^2 - 2
## 4.440892e-16
```

The square root of two cannot be expressed exactly in 16 significant digits. As a result, R has to round the quantity, and the expression resolves to something very close to—but not quite—zero.

These errors are known as *floating-point* errors, and doing arithmetic in these conditions is known as *floating-point arithmetic*. Floating-point arithmetic is not a feature of R; it is a feature of computer programming. Usually floating-point errors won't be enough to ruin your day. Just keep in mind that they may be the cause of surprising results.

You can avoid floating-point errors by avoiding decimals and only using integers. However, this is not an option in most data-science situations. You cannot do much math with integers before you need a noninteger to express the result. Luckily, the errors caused by floating-point arithmetic are usually insignificant (and when they are not, they are easy to spot). As a result, you'll generally use doubles instead of integers as a data scientist.

Characters

A character vector stores small pieces of text. You can create a character vector in R by typing a character or string of characters surrounded by quotes:

```
text <- c("Hello", "World")
text
## "Hello"  "World"

typeof(text)
## "character"

typeof("Hello")
## "character"
```

The individual elements of a character vector are known as *strings*. Note that a string can contain more than just letters. You can assemble a character string from numbers or symbols as well.

> ### Exercise
>
> Can you spot the difference between a character string and a number? Here's a test: Which of these are character strings and which are numbers? 1, "1", "one".

"1" and "one" are both character strings. Character strings can contain number characters, but that doesn't make them numeric. They're just strings that happen to have numbers in them. You can tell strings from real numbers because strings come surrounded by quotes. In fact, anything surrounded by quotes in R will be treated as a character string—no matter what appears between the quotes.

It is easy to confuse R objects with character strings. Why? Because both appear as pieces of text in R code. For example, x is the name of an R object named "x," "x" is a character string that contains the character "x." One is an object that contains raw data, the other is a piece of raw data itself.

Expect an error whenever you forget your quotation marks; R will start looking for an object that probably does not exist.

Logicals

Logical vectors store TRUEs and FALSEs, R's form of Boolean data. Logicals are very helpful for doing things like comparisons:

```
3 > 4
## FALSE
```

Any time you type TRUE or FALSE in capital letters (without quotation marks), R will treat your input as logical data. R also assumes that T and F are shorthand for TRUE and FALSE:

```
logic <- c(TRUE, FALSE, TRUE)
logic
##   TRUE FALSE  TRUE

typeof(logic)
## "logical"

typeof(F)
## "logical"
```

Complex and Raw

Doubles, integers, characters, and logicals are the most common types of atomic vectors in R, but R also recognizes two more types: complex and raw. It is doubtful that you will ever use these to analyze data, but here they are for the sake of thoroughness.

Complex vectors store complex numbers. To create a complex vector, add an imaginary term to a number with i:

```
comp <- c(1 + 1i, 1 + 2i, 1 + 3i)
comp
## 1+1i 1+2i 1+3i

typeof(comp)
## "complex"
```

Raw vectors store raw bytes of data. Making raw vectors gets complicated, but you can make an empty raw vector of length *n* with raw(n). See the help page of raw for more options when working with this type of data:

```
raw(3)
## 00 00 00

typeof(raw(3))
## "raw"
```

> **Exercise**
>
> Create an atomic vector that stores just the face names of the cards in a royal flush, for example, the ace of spades, king of spades, queen of spades, jack of spades, and ten of spades. The face name of the ace of spades would be "ace," and "spades" is the suit.
>
> Which type of vector will you use to save the names?

A character vector is the most appropriate type of atomic vector in which to save card names. You can create one with the c function if you surround each name with quotation marks:

```
hand <- c("ace", "king", "queen", "jack", "ten")
hand
## "ace"   "king"  "queen" "jack"  "ten"

typeof(hand)
## "character"
```

This creates a one-dimensional group of card names—great job! Now let's make a more sophisticated data structure, a two-dimensional table of card names and suits. You can build a more sophisticated object from an atomic vector by giving it some attributes and assigning it a class.

Attributes

An attribute is a piece of information that you can attach to an atomic vector (or any R object). The attribute won't affect any of the values in the object, and it will not appear when you display your object. You can think of an attribute as "metadata"; it is just a convenient place to put information associated with an object. R will normally ignore this metadata, but some R functions will check for specific attributes. These functions may use the attributes to do special things with the data.

You can see which attributes an object has with `attributes`. `attributes` will return NULL if an object has no attributes. An atomic vector, like die, won't have any attributes unless you give it some:

```
attributes(die)
## NULL
```

NULL
R uses NULL to represent the null set, an empty object. NULL is often returned by functions whose values are undefined. You can create a NULL object by typing NULL in capital letters.

Names

The most common attributes to give an atomic vector are names, dimensions (dim), and classes. Each of these attributes has its own helper function that you can use to give attributes to an object. You can also use the helper functions to look up the value of these attributes for objects that already have them. For example, you can look up the value of the names attribute of die with names:

```
names(die)
## NULL
```

NULL means that die does not have a names attribute. You can give one to die by assigning a character vector to the output of names. The vector should include one name for each element in die:

```
names(die) <- c("one", "two", "three", "four", "five", "six")
```

Now die has a names attribute:

```
names(die)
## "one"   "two"   "three" "four"  "five"  "six"

attributes(die)
## $names
## [1] "one"   "two"   "three" "four"  "five"  "six"
```

R will display the names above the elements of die whenever you look at the vector:

```
die
##   one   two three  four  five   six
##     1     2     3     4     5     6
```

However, the names won't affect the actual values of the vector, nor will the names be affected when you manipulate the values of the vector:

```
die + 1
##   one   two three  four  five   six
##     2     3     4     5     6     7
```

You can also use names to change the names attribute or remove it altogether. To change the names, assign a new set of labels to names:

```
names(die) <- c("uno", "dos", "tres", "cuatro", "cinco", "seis")
die
##    uno    dos   tres cuatro  cinco   seis
##      1      2      3      4      5      6
```

To remove the names attribute, set it to NULL:

```
names(die) <- NULL
die
## 1 2 3 4 5 6
```

Dim

You can transform an atomic vector into an *n*-dimensional array by giving it a dimensions attribute with `dim`. To do this, set the `dim` attribute to a numeric vector of length *n*. R will reorganize the elements of the vector into *n* dimensions. Each dimension will have as many rows (or columns, etc.) as the *n*th value of the `dim` vector. For example, you can reorganize `die` into a 2 × 3 matrix (which has 2 rows and 3 columns):

```
dim(die) <- c(2, 3)
die
##      [,1] [,2] [,3]
## [1,]    1    3    5
## [2,]    2    4    6
```

or a 3 × 2 matrix (which has 3 rows and 2 columns):

```
dim(die) <- c(3, 2)
die
##      [,1] [,2]
## [1,]    1    4
## [2,]    2    5
## [3,]    3    6
```

or a 1 × 2 × 3 hypercube (which has 1 row, 2 columns, and 3 "slices"). This is a three-dimensional structure, but R will need to show it slice by slice by slice on your two-dimensional computer screen:

```
dim(die) <- c(1, 2, 3)
die
## , , 1
##
##      [,1] [,2]
## [1,]    1    2
##
## , , 2
##
##      [,1] [,2]
## [1,]    3    4
##
## , , 3
##
##      [,1] [,2]
## [1,]    5    6
```

R will always use the first value in `dim` for the number of rows and the second value for the number of columns. In general, rows always come first in R operations that deal with both rows and columns.

You may notice that you don't have much control over how R reorganizes the values into rows and columns. For example, R always fills up each matrix by columns, instead of by rows. If you'd like more control over this process, you can use one of R's helper functions, matrix or array. They do the same thing as changing the dim attribute, but they provide extra arguments to customize the process.

Matrices

Matrices store values in a two-dimensional array, just like a matrix from linear algebra. To create one, first give matrix an atomic vector to reorganize into a matrix. Then, define how many rows should be in the matrix by setting the nrow argument to a number. matrix will organize your vector of values into a matrix with the specified number of rows. Alternatively, you can set the ncol argument, which tells R how many columns to include in the matrix:

```
m <- matrix(die, nrow = 2)
m
##      [,1] [,2] [,3]
## [1,]    1    3    5
## [2,]    2    4    6
```

matrix will fill up the matrix column by column by default, but you can fill the matrix row by row if you include the argument byrow = TRUE:

```
m <- matrix(die, nrow = 2, byrow = TRUE)
m
##      [,1] [,2] [,3]
## [1,]    1    2    3
## [2,]    4    5    6
```

matrix also has other default arguments that you can use to customize your matrix. You can read about them at matrix's help page (accessible by ?matrix).

Arrays

The array function creates an n-dimensional array. For example, you could use array to sort values into a cube of three dimensions or a hypercube in 4, 5, or *n* dimensions. array is not as customizeable as matrix and basically does the same thing as setting the dim attribute. To use array, provide an atomic vector as the first argument, and a vector of dimensions as the second argument, now called dim:

```
ar <- array(c(11:14, 21:24, 31:34), dim = c(2, 2, 3))
ar
## , , 1
##
##      [,1] [,2]
## [1,]   11   13
## [2,]   12   14
##
## , , 2
##
```

```
##      [,1] [,2]
## [1,]   21   23
## [2,]   22   24
## 
## , , 3
## 
##      [,1] [,2]
## [1,]   31   33
## [2,]   32   34
```

> **Exercise**
>
> Create the following matrix, which stores the name and suit of every card in a royal flush.
>
> ```
> ##      [,1]    [,2]
> ## [1,] "ace"   "spades"
> ## [2,] "king"  "spades"
> ## [3,] "queen" "spades"
> ## [4,] "jack"  "spades"
> ## [5,] "ten"   "spades"
> ```

There is more than one way to build this matrix, but in every case, you will need to start by making a character vector with 10 values. If you start with the following character vector, you can turn it into a matrix with any of the following three commands:

```
hand1 <- c("ace", "king", "queen", "jack", "ten", "spades", "spades",
  "spades", "spades", "spades")

matrix(hand1, nrow = 5)
matrix(hand1, ncol = 2)
dim(hand1) <- c(5, 2)
```

You can also start with a character vector that lists the cards in a slightly different order. In this case, you will need to ask R to fill the matrix row by row instead of column by column:

```
hand2 <- c("ace", "spades", "king", "spades", "queen", "spades", "jack",
  "spades", "ten", "spades")

matrix(hand2, nrow = 5, byrow = TRUE)
matrix(hand2, ncol = 2, byrow = TRUE)
```

Class

Notice that changing the dimensions of your object will not change the type of the object, but it *will* change the object's `class` attribute:

```
dim(die) <- c(2, 3)
typeof(die)
##  "double"

class(die)
##  "matrix"
```

A class is a special case of an atomic vector. For example, the `die` matrix is a special case of a double vector. Every element in the matrix is still a double, but the elements have been arranged into a new structure. R added a `class` attribute to `die` when you changed its dimensions. This class describes `die`'s new format. Many R functions will specifically look for an object's `class` attribute, and then handle the object in a predetermined way based on the attribute.

Note that an object's `class` attribute will not always appear when you run `attributes`; you may need to specifically search for it with `class`:

```
attributes(die)
## $dim
## [1] 2 3
```

You can apply `class` to objects that do not have a `class` attribute. `class` will return a value based on the object's atomic type. Notice that the "class" of a double is "numeric," an odd deviation, but one I am thankful for. I think that the most important property of a double vector is that it contains numbers, a property that "numeric" makes obvious:

```
class("Hello")
##  "character"

class(5)
##  "numeric"
```

You can also use `class` to set an object's `class` attribute, but this is usually a bad idea. R will expect objects of a class to share certain traits, such as attributes, that your object may not possess. You'll learn how to make and use your own classes in Part III.

Dates and Times

The attribute system lets R represent more types of data than just doubles, integers, characters, logicals, complexes, and raws. For example, R uses a special class to represent dates and times. To see this, run **Sys.time()**. Sys.time returns the current time on your computer. The time looks like a character string when you display it, but its data type is actually "double", and its class is "POSIXct" "POSIXt" (it has two classes):

```
now <- Sys.time()
now
## "2014-03-17 12:00:00 UTC"

typeof(now)
##  "double"
```

```
class(now)
## "POSIXct"  "POSIXt"
```

POSIXct is a widely used framework for representing dates and times. In the POSIXct framework, each time is represented by the number of seconds that have passed between the time and 12:00 AM January 1st 1970 (in the Universal Time Coordinated (UTC) zone). For example, the time above occurs 1,395,057,600 seconds after then. So in the POSIXct system, the time would be saved as 1395057600.

R creates the time object by building a double vector with one element, 1395057600. You can see this vector by removing the class attribute of now, or by using the unclass function, which does the same thing:

```
unclass(now)
## 1395057600
```

R then gives the double vector a class attribute that contains two classes, "POSIXct" and "POSIXt". This attribute alerts R functions that they are dealing with a POSIXct time, so they can treat it in a special way. For example, R functions will use the POSIXct standard to convert the time into a user-friendly character string before displaying it.

You can take advantage of this system by giving the POSIXct class to random R objects. For example, have you ever wondered what day it was a million seconds after 12:00 a.m. Jan. 1, 1970?

```
mil <- 1000000
mil
## 1e+06

class(mil) <- c("POSIXct", "POSIXt")
mil
## "1970-01-12 13:46:40 UTC"
```

Jan. 12, 1970. Yikes. A million seconds goes by faster than you would think. This conversion worked well because the POSIXct class does not rely on any additional attributes, but in general, forcing the class of an object is a bad idea.

There are many different classes of data in R and its packages, and new classes are invented every day. It would be difficult to learn about every class, but you do not have to. Most classes are only useful in specific situations. Since each class comes with its own help page, you can wait to learn about a class until you encounter it. However, there is one class of data that is so ubiquitous in R that you should learn about it alongside the atomic data types. That class is factors.

Factors

Factors are R's way of storing categorical information, like ethnicity or eye color. Think of a factor as something like a gender; it can only have certain values (male or female), and these values may have their own idiosyncratic order (ladies first). This arrangement makes factors very useful for recording the treatment levels of a study and other categorical variables.

To make a factor, pass an atomic vector into the `factor` function. R will recode the data in the vector as integers and store the results in an integer vector. R will also add a `levels` attribute to the integer, which contains a set of labels for displaying the factor values, and a `class` attribute, which contains the class `factor`:

```
gender <- factor(c("male", "female", "female", "male"))

typeof(gender)
## "integer"

attributes(gender)
## $levels
## [1] "female" "male"
##
## $class
## [1] "factor"
```

You can see exactly how R is storing your factor with `unclass`:

```
unclass(gender)
## [1] 2 1 1 2
## attr(,"levels")
## [1] "female" "male"
```

R uses the levels attribute when it displays the factor, as you will see. R will display each 1 as `female`, the first label in the levels vector, and each 2 as `male`, the second label. If the factor included 3s, they would be displayed as the third label, and so on:

```
gender
## male   female female male
## Levels: female male
```

Factors make it easy to put categorical variables into a statistical model because the variables are already coded as numbers. However, factors can be confusing since they look like character strings but behave like integers.

R will often try to convert character strings to factors when you load and create data. In general, you will have a smoother experience if you do not let R make factors until you ask for them. I'll show you how to do this when we start reading in data.

You can convert a factor to a character string with the `as.character` function. R will retain the display version of the factor, not the integers stored in memory:

```
as.character(gender)
## "male"   "female" "female" "male"
```

Now that you understand the possibilities provided by R's atomic vectors, let's make a more complicated type of playing card.

> ### Exercise
>
> Many card games assign a numerical value to each card. For example, in blackjack, each face card is worth 10 points, each number card is worth between 2 and 10 points, and each ace is worth 1 or 11 points, depending on the final score.
>
> Make a virtual playing card by combining "ace," "heart," and 1 into a vector. What type of atomic vector will result? Check if you are right.

You may have guessed that this exercise would not go well. Each atomic vector can only store one type of data. As a result, R coerces all of your values to character strings:

```
card <- c("ace", "hearts", 1)
card
## "ace"    "hearts" "1"
```

This will cause trouble if you want to do math with that point value, for example, to see who won your game of blackjack.

Data types in vectors

If you try to put multiple types of data into a vector, R will convert the elements to a single type of data.

Since matrices and arrays are special cases of atomic vectors, they suffer from the same behavior. Each can only store one type of data.

This creates a couple of problems. First, many data sets contain multiple types of data. Simple programs like Excel and Numbers can save multiple types of data in the same data set, and you should hope that R can too. Don't worry, it can.

Second, coercion is a common behavior in R, so you'll want to know how it works.

Coercion

R's coercion behavior may seem inconvenient, but it is not arbitrary. R always follows the same rules when it coerces data types. Once you are familiar with these rules, you can use R's coercion behavior to do surprisingly useful things.

So how does R coerce data types? If a character string is present in an atomic vector, R will convert everything else in the vector to character strings. If a vector only contains logicals and numbers, R will convert the logicals to numbers; every TRUE becomes a 1, and every FALSE becomes a 0, as shown in Figure 3-1.

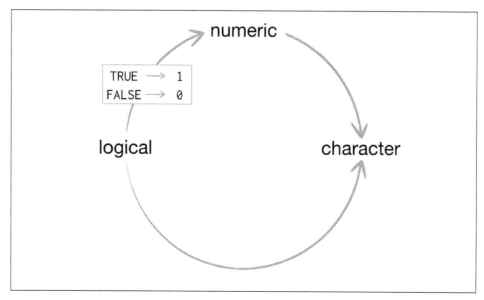

Figure 3-1. R always uses the same rules to coerce data to a single type. If character strings are present, everything will be coerced to a character string. Otherwise, logicals are coerced to numerics.

This arrangement preserves information. It is easy to look at a character string and tell what information it used to contain. For example, you can easily spot the origins of "TRUE" and "5". You can also easily back-transform a vector of 1s and 0s to TRUEs and FALSEs.

R uses the same coercion rules when you try to do math with logical values. So the following code:

```
sum(c(TRUE, TRUE, FALSE, FALSE))
```

will become:

```
sum(c(1, 1, 0, 0))
## 2
```

This means that sum will count the number of TRUEs in a logical vector (and mean will calculate the proportion of TRUEs). Neat, huh?

You can explicitly ask R to convert data from one type to another with the as functions. R will convert the data whenever there is a sensible way to do so:

```
as.character(1)
## "1"

as.logical(1)
## TRUE

as.numeric(FALSE)
## 0
```

You now know how R coerces data types, but this won't help you save a playing card. To do that, you will need to avoid coercion altogether. You can do this by using a new type of object, a *list*.

Before we look at lists, let's address a question that might be on your mind.

Many data sets contain multiple types of information. The inability of vectors, matrices, and arrays to store multiple data types seems like a major limitation. So why bother with them?

In some cases, using only a single type of data is a huge advantage. Vectors, matrices, and arrays make it very easy to do math on large sets of numbers because R knows that it can manipulate each value the same way. Operations with vectors, matrices, and arrays also tend to be fast because the objects are so simple to store in memory.

In other cases, allowing only a single type of data is not a disadvantage. Vectors are the most common data structure in R because they store variables very well. Each value in a variable measures the same property, so there's no need to use different types of data.

Lists

Lists are like atomic vectors because they group data into a one-dimensional set. However, lists do not group together individual values; lists group together R objects, such as atomic vectors and other lists. For example, you can make a list that contains a numeric vector of length 31 in its first element, a character vector of length 1 in its second element, and a new list of length 2 in its third element. To do this, use the `list` function.

`list` creates a list the same way `c` creates a vector. Separate each element in the list with a comma:

```
list1 <- list(100:130, "R", list(TRUE, FALSE))
list1
## [[1]]
## [1] 100 101 102 103 104 105 106 107 108 109 110 111 112
## [14] 113 114 115 116 117 118 119 120 121 122 123 124 125
## [27] 126 127 128 129 130
##
## [[2]]
## [1] "R"
##
```

```
## [[3]]
## [[3]][[1]]
## [1] TRUE
##
## [[3]][[2]]
## [1] FALSE
```

I left the [1] notation in the output so you can see how it changes for lists. The double-bracketed indexes tell you which element of the list is being displayed. The single-bracket indexes tell you which subelement of an element is being displayed. For example, 100 is the first subelement of the first element in the list. "R" is the first sub-element of the second element. This two-system notation arises because each element of a list can be *any* R object, including a new vector (or list) with its own indexes.

Lists are a basic type of object in R, on par with atomic vectors. Like atomic vectors, they are used as building blocks to create many more spohisticated types of R objects.

As you can imagine, the structure of lists can become quite complicated, but this flexibility makes lists a useful all-purpose storage tool in R: you can group together anything with a list.

However, not every list needs to be complicated. You can store a playing card in a very simple list.

> **Exercise**
>
> Use a list to store a single playing card, like the ace of hearts, which has a point value of one. The list should save the face of the card, the suit, and the point value in separate elements.

You can create your card like this. In the following example, the first element of the list is a character vector (of length 1). The second element is also a character vector, and the third element is a numeric vector:

```
card <- list("ace", "hearts", 1)
card
## [[1]]
## [1] "ace"
##
## [[2]]
## [1] "hearts"
##
## [[3]]
## [1] 1
```

You can also use a list to store a whole deck of playing cards. Since you can save a single playing card as a list, you can save a deck of playing cards as a list of 52 sublists (one for each card). But let's not bother—there's a much cleaner way to do the same thing. You can use a special class of list, known as a *data frame*.

Data Frames

Data frames are the two-dimensional version of a list. They are far and away the most useful storage structure for data analysis, and they provide an ideal way to store an entire deck of cards. You can think of a data frame as R's equivalent to the Excel spreadsheet because it stores data in a similar format.

Data frames group vectors together into a two-dimensional table. Each vector becomes a column in the table. As a result, each column of a data frame can contain a different type of data; but within a column, every cell must be the same type of data, as in Figure 3-2.

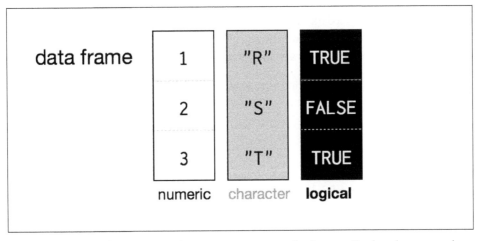

Figure 3-2. Data frames store data as a sequence of columns. Each column can be a different data type. Every column in a data frame must be the same length.

Creating a data frame by hand takes a lot of typing, but you can do it (if you like) with the data.frame function. Give data.frame any number of vectors, each separated with a comma. Each vector should be set equal to a name that describes the vector. data.frame will turn each vector into a column of the new data frame:

```
df <- data.frame(face = c("ace", "two", "six"),
  suit = c("clubs", "clubs", "clubs"), value = c(1, 2, 3))
df
## face  suit value
##  ace clubs     1
##  two clubs     2
##  six clubs     3
```

You'll need to make sure that each vector is the same length (or can be made so with R's recycling rules; see Figure 1-4), as data frames cannot combine columns of different lengths.

In the previous code, I named the arguments in data.frame face, suit, and value, but you can name the arguments whatever you like. data.frame will use your argument names to label the columns of the data frame.

Names

You can also give names to a list or vector when you create one of these objects. Use the same syntax as with data.frame:

```
list(face = "ace", suit = "hearts", value = 1)
c(face = "ace", suit = "hearts", value = "one")
```

The names will be stored in the object's names attribute.

If you look at the type of a data frame, you will see that it is a list. In fact, each data frame is a list with class data.frame. You can see what types of objects are grouped together by a list (or data frame) with the str function:

```
typeof(df)
## "list"

class(df)
## "data.frame"

str(df)
## 'data.frame':    3 obs. of  3 variables:
##  $ face : Factor w/ 3 levels "ace","six","two": 1 3 2
##  $ suit : Factor w/ 1 level "clubs": 1 1 1
##  $ value: num  1 2 3
```

Notice that R saved your character strings as factors. I told you that R likes factors! It is not a very big deal here, but you can prevent this behavior by adding the argument stringsAsFactors = FALSE to data.frame:

```
df <- data.frame(face = c("ace", "two", "six"),
  suit = c("clubs", "clubs", "clubs"), value = c(1, 2, 3),
  stringsAsFactors = FALSE)
```

A data frame is a great way to build an entire deck of cards. You can make each row in the data frame a playing card, and each column a type of value—each with its own appropriate data type. The data frame would look something like this:

```
##   face  suit value
##   king spades   13
##  queen spades   12
##   jack spades   11
##    ten spades   10
```

```
##     nine   spades    9
##    eight   spades    8
##    seven   spades    7
##      six   spades    6
##     five   spades    5
##     four   spades    4
##    three   spades    3
##      two   spades    2
##      ace   spades    1
##     king    clubs   13
##    queen    clubs   12
##     jack    clubs   11
##      ten    clubs   10
##    ... and so on.
```

You could create this data frame with `data.frame`, but look at the typing involved! You need to write three vectors, each with 52 elements:

```
deck <- data.frame(
  face = c("king", "queen", "jack", "ten", "nine", "eight", "seven", "six",
    "five", "four", "three", "two", "ace", "king", "queen", "jack", "ten",
    "nine", "eight", "seven", "six", "five", "four", "three", "two", "ace",
    "king", "queen", "jack", "ten", "nine", "eight", "seven", "six", "five",
    "four", "three", "two", "ace", "king", "queen", "jack", "ten", "nine",
    "eight", "seven", "six", "five", "four", "three", "two", "ace"),
  suit = c("spades", "spades", "spades", "spades", "spades", "spades",
    "spades", "spades", "spades", "spades", "spades", "spades", "spades",
    "clubs", "clubs", "clubs", "clubs", "clubs", "clubs", "clubs", "clubs",
    "clubs", "clubs", "clubs", "clubs", "clubs", "diamonds", "diamonds",
    "diamonds", "diamonds", "diamonds", "diamonds", "diamonds", "diamonds",
    "diamonds", "diamonds", "diamonds", "diamonds", "diamonds", "hearts",
    "hearts", "hearts", "hearts", "hearts", "hearts", "hearts", "hearts",
    "hearts", "hearts", "hearts", "hearts", "hearts"),
  value = c(13, 12, 11, 10, 9, 8, 7, 6, 5, 4, 3, 2, 1, 13, 12, 11, 10, 9, 8,
    7, 6, 5, 4, 3, 2, 1, 13, 12, 11, 10, 9, 8, 7, 6, 5, 4, 3, 2, 1, 13, 12, 11,
    10, 9, 8, 7, 6, 5, 4, 3, 2, 1)
)
```

You should avoid typing large data sets in by hand whenever possible. Typing invites typos and errors, not to mention RSI. It is always better to acquire large data sets as a computer file. You can then ask R to read the file and store the contents as an object.

I've created a file for you to load that contains a data frame of playing-card information, so don't worry about typing in the code. Instead, turn your attention toward loading data into R.

Loading Data

You can load the deck data frame from the file *deck.csv* (*http://bit.ly/deck_CSV*). Please take a moment to download the file before reading on. Visit the website, click "Download Gist," and then open the folder that your web browser downloads. *deck.csv* will be inside.

deck.csv is a comma-separated values file, or CSV for short. CSVs are plain-text files, which means you can open them in a text editor (as well as many other programs). If you open *desk.csv*, you'll notice that it contains a table of data that looks like the following table. Each row of the table is saved on its own line, and a comma is used to separate the cells within each row. Every CSV file shares this basic format:

```
"face","suit,"value"
"king","spades",13
"queen","spades,12
"jack","spades,11
"ten","spades,10
"nine","spades,9
... and so on.
```

Most data-science applications can open plain-text files and export data as plain-text files. This makes plain-text files a sort of lingua franca for data science.

To load a plain-text file into R, click the Import Dataset icon in RStudio, shown in Figure 3-3. Then select "From text file."

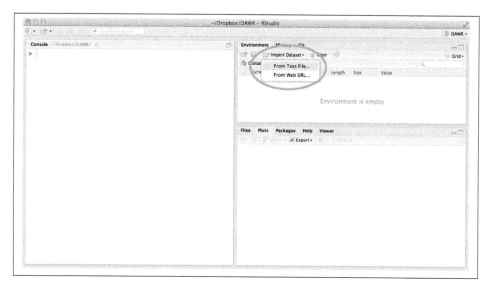

Figure 3-3. You can import data from plain-text files with RStudio's Import Dataset.

RStudio will ask you to select the file you want to import, then it will open a wizard to help you import the data, as in Figure 3-4. Use the wizard to tell RStudio what name to give the data set. You can also use the wizard to tell RStudio which character the data set uses as a separator, which character it uses to represent decimals (usually a period in the United States and a comma in Europe), and whether or not the data set comes with a row of column names (known as a *header*). To help you out, the wizard shows you what the raw file looks like, as well as what your loaded data will look like based on the input settings.

You can also unclick the box "Strings as factors" in the wizard. I recommend doing this. If you do, R will load all of your character strings as character strings. If you do not, R will convert them to factors.

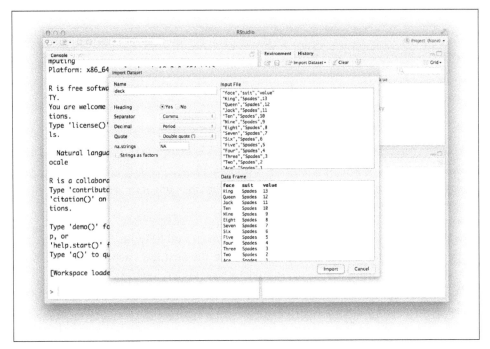

Figure 3-4. RStudio's import wizard.

Once everything looks right, click Import. RStudio will read in the data and save it to a data frame. RStudio will also open a data viewer, so you can see your new data in a spreadsheet format. This is a good way to check that everything came through as expected. If all worked well, your file should appear in a View tab of RStudio, like in Figure 3-5. You can examine the data frame in the console with **head(deck)**.

Online data
You can load a plain-text file straight from the Internet by clicking the "From Web URL…" option under Import Dataset. The file will need to have its own URL, and you will need to be connected.

Loading Data | 59

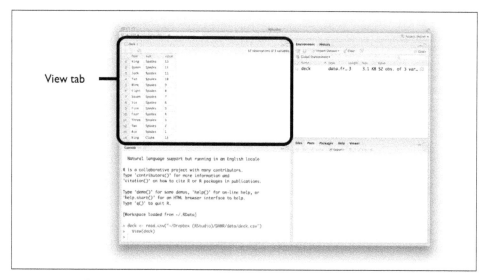

Figure 3-5. When you import a data set, RStudio will save the data to a data frame and then display the data frame in a View tab. You can open any data frame in a View tab at any time with the View function.

Now it is your turn. Download *deck.csv* and import it into RStudio. Be sure to save the output to an R object called deck: you'll use it in the next few chapters. If everything goes correctly, the first few lines of your data frame should look like this:

```
head(deck) ❶
##   face  suit value
##   king spades    13
##  queen spades    12
##   jack spades    11
##    ten spades    10
##   nine spades     9
##  eight spades     8
```

❶ head and `tail` are two functions that provide an easy way to peek at large data sets. head will return just the first six rows of the data set, and `tail` will return just the last six rows. To see a different number of rows, give head or `tails` a second argument, the number of rows you would like to view, for example, head(deck, 10).

R can open many types of files—not just CSVs. Visit Appendix D to learn how to open other common types of files in R.

Saving Data

Before we go any further, let's save a copy of deck as a new *.csv* file. That way you can email it to a colleague, store it on a thumb drive, or open it in a different program. You can save any data frame in R to a *.csv* file with the command write.csv. To save deck, run:

    write.csv(deck, file = "cards.csv", row.names = FALSE)

R will turn your data frame into a plain-text file with the comma-separated values format and save the file to your working directory. To see where your working directory is, run **getwd()**. To change the location of your working directory, visit Session > Set Working Directory > Choose Directory in the RStudio menu bar.

You can customize the save process with write.csv's large set of optional arguments (see ?write.csv for details). However, there are three arguments that you should use *every* time you run write.csv.

First, you should give write.csv the name of the data frame that you wish to save. Next, you should provide a file name to give your file. R will take this name quite literally, so be sure to provide an extension.

Finally, you should add the argument row.names = FALSE. This will prevent R from adding a column of numbers at the start of your data frame. These numbers will identify your rows from 1 to 52, but it is unlikely that whatever program you open *cards.csv* in will understand the row name system. More than likely, the program will assume that the row names are the first column of data in your data frame. In fact, this is exactly what R will assume if you reopen *cards.csv*. If you save and open *cards.csv* several times in R, you'll notice duplicate columns of row numbers forming at the start of your data frame. I can't explain why R does this, but I can explain how to avoid it: use row.names = FALSE whenever you save data with write.csv.

For more details about saving files, including how to compress saved files and how to save files in other formats, see Appendix D.

Good work. You now have a virtual deck of cards to work with. Take a breather, and when you come back, we'll start writing some functions to use on your deck.

Summary

You can save data in R with five different objects, which let you store different types of values in different types of relationships, as in Figure 3-6. Of these objects, data frames are by far the most useful for data science. Data frames store one of the most common forms of data used in data science, tabular data.

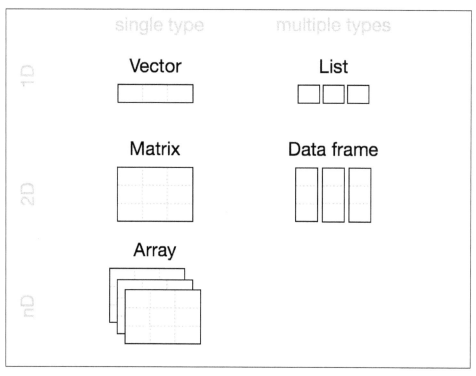

Figure 3-6. R's most common data structures are vectors, matrices, arrays, lists, and data frames.

You can load tabular data into a data frame with RStudio's Import Dataset button—so long as the data is saved as a plain-text file. This requirement is not as limiting as it sounds. Most software programs can export data as a plain-text file. So if you have an Excel file (for example) you can open the file in Excel and export the data as a CSV to use with R. In fact, opening a file in its original program is good practice. Excel files use metadata, like sheets and formulas, that help Excel work with the file. R can try to extract raw data from the file, but it won't be as good at doing this as Microsoft Excel is. No program is better at converting Excel files than Excel. Similarly, no program is better at converting SAS Xport files than SAS, and so on.

However, you may find yourself with a program-specific file, but not the program that created it. You wouldn't want to buy a multi-thousand-dollar SAS license just to open a SAS file. Thankfully R *can* open many types of files, including files from other programs and databases. R even has its own program-specific formats that can help you save memory and time if you know that you will be working entirely in R. If you'd like to know more about all of your options for loading and saving data in R, see Appendix D.

Chapter 4 will build upon the skills you learned in this chapter. Here, you learned how to store data in R. In Chapter 4, you will learn how to access values once they've been stored. You'll also write two functions that will let you start using your deck, a shuffle function and a deal function.

CHAPTER 4
R Notation

Now that you have a deck of cards, you need a way to do card-like things with it. First, you'll want to reshuffle the deck from time to time. And next, you'll want to deal cards from the deck (one card at a time, whatever card is on top—we're not cheaters).

To do these things, you'll need to work with the individual values inside your data frame, a task essential to data science. For example, to deal a card from the top of your deck, you'll need to write a function that selects the first row of values in your data frame, like this:

```
deal(deck)
##  face   suit value
##  king spades   13
```

You can select values within an R object with R's notation system.

Selecting Values

R has a notation system that lets you extract values from R objects. To extract a value or set of values from a data frame, write the data frame's name followed by a pair of hard brackets:

```
deck[ , ]
```

Between the brackets will go two indexes separated by a comma. The indexes tell R which values to return. R will use the first index to subset the rows of the data frame and the second index to subset the columns.

You have a choice when it comes to writing indexes. There are six different ways to write an index for R, and each does something slightly different. They are all very simple and quite handy, so let's take a look at each of them. You can create indexes with:

- Positive integers
- Negative integers
- Zero
- Blank spaces
- Logical values
- Names

The simplest of these to use is positive integers.

Positive Integers

R treats positive integers just like *ij* notation in linear algebra: `deck[i,j]` will return the value of `deck` that is in the *ith* row and the *jth* column, Figure 4-1. Notice that *i* and *j* only need to be integers in the mathematical sense. They can be saved as numerics in R:

```
head(deck)
##    face    suit value
##    king  spades    13
##   queen  spades    12
##    jack  spades    11
##     ten  spades    10
##    nine  spades     9
##   eight  spades     8

deck[1, 1]
## "king"
```

To extract more than one value, use a vector of positive integers. For example, you can return the first row of `deck` with `deck[1, c(1, 2, 3)]` or `deck[1, 1:3]`:

```
deck[1, c(1, 2, 3)]
## face   suit value
## king spades    13
```

R will return the values of `deck` that are in both the first row and the first, second, and third columns. Note that R won't actually remove these values from `deck`. R will give you a new set of values which are copies of the original values. You can then save this new set to an R object with R's assignment operator:

```
new <- deck[1, c(1, 2, 3)]
new
## face   suit value
## king spades    13
```

Repetition
If you repeat a number in your index, R will return the corresponding value(s) more than once in your "subset." This code will return the first row of deck twice:

```
deck[c(1, 1), c(1, 2, 3)]
##       face   suit value
## king       spades    13
## king       spades    13
```

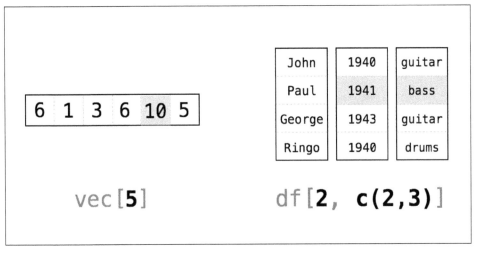

Figure 4-1. R uses the ij notation system of linear algebra. The commands in this figure will return the shaded values.

R's notation system is not limited to data frames. You can use the same syntax to select values in any R object, as long as you supply one index for each dimension of the object. So, for example, you can subset a vector (which has one dimension) with a single index:

```
vec <- c(6, 1, 3, 6, 10, 5)
vec[1:3]
## 6 1 3
```

Selecting Values | 67

Indexing begins at 1

In some programming languages, indexing begins with 0. This means that 0 returns the first element of a vector, 1 returns the second element, and so on.

This isn't the case with R. Indexing in R behaves just like indexing in linear algebra. The first element is always indexed by 1. Why is R different? Maybe because it was written for mathematicians. Those of us who learned indexing from a linear algebra course wonder why computers programmers start with 0.

drop = FALSE

If you select two or more columns from a data frame, R will return a new data frame:

```
deck[1:2, 1:2]
##   face    suit
##   king  spades
##  queen  spades
```

However, if you select a single column, R will return a vector:

```
deck[1:2, 1]
## "king"  "queen"
```

If you would prefer a data frame instead, you can add the optional argument drop = FALSE between the brackets:

```
deck[1:2, 1, drop = FALSE]
##   face
##   king
##  queen
```

This method also works for selecting a single column from a matrix or an array.

Negative Integers

Negative integers do the exact opposite of positive integers when indexing. R will return every element *except* the elements in a negative index. For example, deck[-1, 1:3] will return everything *but* the first row of deck. deck[-(2:52), 1:3] will return the first row (and exclude everything else):

```
deck[-(2:52), 1:3]
## face    suit  value
## king  spades    13
```

Negative integers are a more efficient way to subset than positive integers if you want to include the majority of a data frame's rows or columns.

R will return an error if you try to pair a negative integer with a positive integer in the *same* index:

```
deck[c(-1, 1), 1]
## Error in xj[i] : only 0's may be mixed with negative subscripts
```

However, you can use both negative and positive integers to subset an object if you use them in *different* indexes (e.g., if you use one in the rows index and one in the columns index, like deck[-1, 1]).

Zero

What would happen if you used zero as an index? Zero is neither a positive integer nor a negative integer, but R will still use it to do a type of subsetting. R will return nothing from a dimension when you use zero as an index. This creates an empty object:

```
deck[0, 0]
## data frame with 0 columns and 0 rows
```

To be honest, indexing with zero is not very helpful.

Blank Spaces

You can use a blank space to tell R to extract *every* value in a dimension. This lets you subset an object on one dimension but not the others, which is useful for extracting entire rows or columns from a data frame:

```
deck[1, ]
##   face  suit value
## king spades    13
```

Logical Values

If you supply a vector of TRUEs and FALSEs as your index, R will match each TRUE and FALSE to a row in your data frame (or a column depending on where you place the index). R will then return each row that corresponds to a TRUE, Figure 4-2.

It may help to imagine R reading through the data frame and asking, "Should I return the *i*th row of the data structure?" and then consulting the *i*th value of the index for its answer. For this system to work, your vector must be as long as the dimension you are trying to subset:

```
deck[1, c(TRUE, TRUE, FALSE)]
##   face  suit
## king spades

rows <- c(TRUE, F, F, F, F, F, F, F, F, F, F, F, F, F, F, F,
    F, F, F, F, F, F, F, F, F, F, F, F, F, F, F, F, F,
    F, F, F, F, F, F, F, F, F, F, F, F, F, F, F)
deck[rows, ]
##   face  suit value
## king spades    13
```

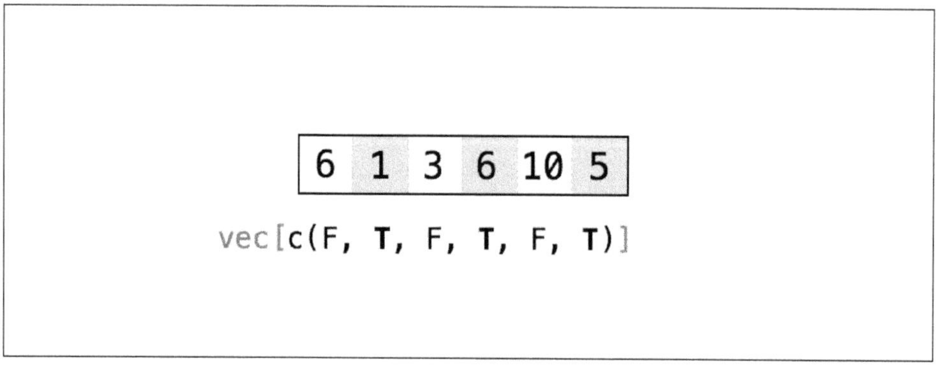

```
       ┌─────────────────────┐
       │ 6   1   3   6  10  5│
       └─────────────────────┘
        vec[c(F, T, F, T, F, T)]
```

Figure 4-2. You can use vectors of TRUEs and FALSEs to tell R exactly which values you want to extract and which you do not. The command would return just the numbers 1, 6, and 5.

This system may seem odd—who wants to type so many TRUEs and FALSEs?—but it will become very powerful in Chapter 5.

Names

Finally, you can ask for the elements you want by name—if your object has names (see "Names" on page 44). This is a common way to extract the columns of a data frame, since columns almost always have names:

```
deck[1, c("face", "suit", "value")]
## face    suit value
## king spades    13

# the entire value column
deck[ , "value"]
##  13 12 11 10  9  8  7  6  5  4  3  2  1 13 12 11 10  9  8
##   7  6  5  4  3  2  1 13 12 11 10  9  8  7  6  5  4  3  2
##   1 13 12 11 10  9  8  7  6  5  4  3  2  1
```

Deal a Card

Now that you know the basics of R's notation system, let's put it to use.

Exercise

Complete the following code to make a function that returns the first row of a data frame:

```
deal <- function(cards) {
  # ?
}
```

You can use any of the systems to write a `deal` function that returns the first row of your data frame. I'll use positive integers and blanks because I think they are easy to understand:

```
deal <- function(cards) {
  cards[1, ]
}
```

The function does exactly what you want: it deals the top card from your data set. However, the function becomes less impressive if you run `deal` over and over again:

```
deal(deck)
## face    suit value
## king spades    13

deal(deck)
## face    suit value
## king spades    13

deal(deck)
## face    suit value
## king spades    13
```

`deal` always returns the king of spades because `deck` doesn't know that we've dealt the card away. Hence, the king of spades stays where it is, at the top of the deck ready to be dealt again. This is a difficult problem to solve, and we will *deal* with it in Chapter 6. In the meantime, you can fix the problem by shuffling your deck after every deal. Then a new card will always be at the top.

Shuffling is a temporary compromise: the probabilities at play in your deck will not match the probabilities that occur when you play a game with a single deck of cards. For example, there will still be a probability that the king of spades appears twice in a row. However, things are not as bad as they may seem. Most casinos use five or six decks at a time in card games to prevent card counting. The probabilities that you would encounter in those situations are very close to the ones we will create here.

Shuffle the Deck

When you shuffle a real deck of cards, you randomly rearrange the order of the cards. In your virtual deck, each card is a row in a data frame. To shuffle the deck, you need to randomly reorder the rows in the data frame. Can this be done? You bet! And you already know everything you need to do it.

This may sound silly, but start by extracting every row in your data frame:

```
deck2 <- deck[1:52, ]

head(deck2)
##    face   suit value
##    king spades    13
##   queen spades    12
##    jack spades    11
##     ten spades    10
##    nine spades     9
##   eight spades     8
```

What do you get? A new data frame whose order hasn't changed at all. What if you asked R to extract the rows in a different order? For example, you could ask for row 2, *then* row 1, and then the rest of the cards:

```
deck3 <- deck[c(2, 1, 3:52), ]

head(deck3)
##    face   suit value
##   queen spades    12
##    king spades    13
##    jack spades    11
##     ten spades    10
##    nine spades     9
##   eight spades     8
```

R complies. You'll get all the rows back, and they'll come in the order you ask for them. If you want the rows to come in a random order, then you need to sort the integers from 1 to 52 into a random order and use the results as a row index. How could you generate such a random collection of integers? With our friendly neighborhood `sample` function:

```
random <- sample(1:52, size = 52)
random
## 35 28 39  9 18 29 26 45 47 48 23 22 21 16 32 38  1 15 20
## 11  2  4 14 49 34 25  8  6 10 41 46 17 33  5  7 44  3 27
## 50 12 51 40 52 24 19 13 42 37 43 36 31 30

deck4 <- deck[random, ]
head(deck4)
##    face     suit value
##    five diamonds     5
##   queen diamonds    12
##     ace diamonds     1
##    five   spades     5
##    nine    clubs     9
##    jack diamonds    11
```

Now the new set is truly shuffled. You'll be finished once you wrap these steps into a function.

> **Exercise**
>
> Use the preceding ideas to write a `shuffle` function. `shuffle` should take a data frame and return a shuffled copy of the data frame.

Your `shuffle` function will look like the one that follows:

```
shuffle <- function(cards) { 
  random <- sample(1:52, size = 52)
  cards[random, ]
}
```

Nice work! Now you can shuffle your cards between each deal:

```
deal(deck)
## face   suit value
## king spades    13

deck2 <- shuffle(deck)

deal(deck2)
## face   suit value
## jack  clubs    11
```

Dollar Signs and Double Brackets

Two types of object in R obey an optional second system of notation. You can extract values from data frames and lists with the $ syntax. You will encounter the $ syntax again and again as an R programmer, so let's examine how it works.

To select a column from a data frame, write the data frame's name and the column name separated by a $. Notice that no quotes should go around the column name:

```
deck$value
## 13 12 11 10  9  8  7  6  5  4  3  2  1 13 12 11 10  9  8  7
##  6  5  4  3  2  1 13 12 11 10  9  8  7  6  5  4  3  2  1 13
## 12 11 10  9  8  7  6  5  4  3  2  1
```

R will return all of the values in the column as a vector. This $ notation is incredibly useful because you will often store the variables of your data sets as columns in a data frame. From time to time, you'll want to run a function like `mean` or `median` on the values in a variable. In R, these functions expect a vector of values as input, and `deck$value` delivers your data in just the right format:

```
mean(deck$value)
## 7

median(deck$value)
## 7
```

You can use the same $ notation with the elements of a list, if they have names. This notation has an advantage with lists, too. If you subset a list in the usual way, R will return a *new* list that has the elements you requested. This is true even if you only request a single element.

To see this, make a list:

```
lst <- list(numbers = c(1, 2), logical = TRUE, strings = c("a", "b", "c"))
lst
## $numbers
## [1] 1 2

## $logical
## [1] TRUE

## $strings
## [1] "a" "b" "c"
```

And then subset it:

```
lst[1]
## $numbers
## [1] 1 2
```

The result is a smaller *list* with one element. That element is the vector c(1, 2). This can be annoying because many R functions do not work with lists. For example, sum(lst[1]) will return an error. It would be horrible if once you stored a vector in a list, you could only ever get it back as a list:

```
sum(lst[1])
## Error in sum(lst[1]) : invalid 'type' (list) of argument
```

When you use the $ notation, R will return the selected values as they are, with no list structure around them:

```
lst$numbers
## 1 2
```

You can then immediately feed the results to a function:

```
sum(lst$numbers)
## 3
```

If the elements in your list do not have names (or you do not wish to use the names), you can use two brackets, instead of one, to subset the list. This notation will do the same thing as the $ notation:

```
lst[[1]]
## 1 2
```

In other words, if you subset a list with single-bracket notation, R will return a smaller list. If you subset a list with double-bracket notation, R will return just the values that were inside an element of the list. You can combine this feature with any of R's indexing methods:

```
lst["numbers"]
## $numbers
## [1] 1 2

lst[["numbers"]]
## 1 2
```

This difference is subtle but important. In the R community, there is a popular, and helpful, way to think about it, Figure 4-3. Imagine that each list is a train and each element is a train car. When you use single brackets, R selects individual train cars and returns them as a new train. Each car keeps its contents, but those contents are still inside a train car (i.e., a list). When you use double brackets, R actually unloads the car and gives you back the contents.

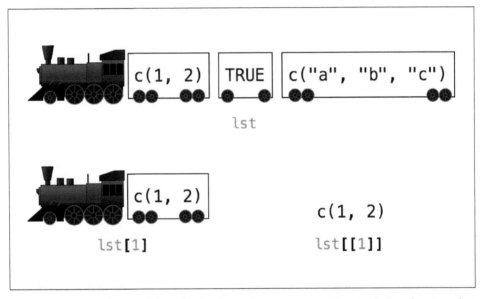

Figure 4-3. It can be helpful to think of your list as a train. Use single brackets to select train cars, double brackets to select the contents inside of a car.

Never attach

In R's early days, it became popular to use attach() on a data set once you had it loaded. Don't do this! attach recreates a computing environment similar to those used in other statistics applications like Stata and SPSS, which crossover users liked. However, R is not Stata or SPSS. R is optimized to use the R computing environment, and running attach() can cause confusion with some R functions.

What does attach() do? On the surface, attach saves you typing. If you attach the deck data set, you can refer to each of its variables by name; instead of typing deck$face, you can just type face. But typing isn't bad. It gives you a chance to be explicit, and in computer programming, explicit is good. Attaching a data set creates the possibility that R will confuse two variable names. If this occurs within a function, you're likely to get unusable results and an unhelpful error message to explain what happened.

Now that you are an expert at retrieving values stored in R, let's summarize what you've accomplished.

Summary

You have learned how to access values that have been stored in R. You can retrieve a copy of values that live inside a data frame and use the copies for new computations.

In fact, you can use R's notation system to access values in any R object. To use it, write the name of an object followed by brackets and indexes. If your object is one-dimensional, like a vector, you only need to supply one index. If it is two-dimensional, like a data frame, you need to supply two indexes separated by a comma. And, if it is n-dimensional, you need to supply n indexes, each separated by a comma.

In Chapter 5, you'll take this system a step further and learn how to change the actual values that are stored inside your data frame. This is all adding up to something special: complete control of your data. You can now store your data in your computer, retrieve individual values at will, and use your computer to perform correct calculations with those values.

Does this sound basic? It may be, but it is also powerful and essential for efficient data science. You no longer need to memorize everything in your head, nor worry about doing mental arithmetic wrong. This low-level control over your data is also a prerequisite for more efficient R programs, the subject of Part III.

CHAPTER 5
Modifying Values

Are you ready to play some games with your virtual deck? Not so fast! The point system in your deck of cards doesn't align well with many card games. For example, in war and poker, aces are usually scored higher than kings. They'd have a point value of 14, not 1.

In this task, you will change the point system of your deck three times to match three different games: war, hearts, and blackjack. Each of these games will teach you something different about modifying the values inside of a data set. Start by making a copy of deck that you can manipulate. This will ensure that you always have a pristine copy of deck to fall back on (should things go awry):

```
deck2 <- deck
```

Changing Values in Place

You can use R's notation system to modify values within an R object. First, describe the value (or values) you wish to modify. Then use the assignment operator <- to overwrite those values. R will update the selected values *in the original object*. Let's put this into action with a real example:

```
vec <- c(0, 0, 0, 0, 0, 0)
vec
##  0 0 0 0 0 0
```

Here's how you can select the first value of vec:

```
vec[1]
##  0
```

And here is how you can modify it:

```
vec[1] <- 1000
vec
##  1000    0    0    0    0    0
```

77

You can replace multiple values at once as long as the number of new values equals the number of selected values:

```
vec[c(1, 3, 5)] <- c(1, 1, 1)
vec
## 1 0 1 0 1 0

vec[4:6] <- vec[4:6] + 1
vec
## 1 0 1 1 2 1
```

You can also create values that do not yet exist in your object. R will expand the object to accommodate the new values:

```
vec[7] <- 0
vec
## 1 0 1 1 2 1 0
```

This provides a great way to add new variables to your data set:

```
deck2$new <- 1:52

head(deck2)
##     face   suit value new
##     king spades    13   1
##    queen spades    12   2
##     jack spades    11   3
##      ten spades    10   4
##     nine spades     9   5
##    eight spades     8   6
```

You can also remove columns from a data frame (and elements from a list) by assigning them the symbol NULL:

```
deck2$new <- NULL

head(deck2)
##     face   suit value
##     king spades    13
##    queen spades    12
##     jack spades    11
##      ten spades    10
##     nine spades     9
##    eight spades     8
```

In the game of war, aces are king (figuratively speaking). They receive the highest value of all the cards, which would be something like 14. Every other card gets the value that it already has in deck. To play war, you just need to change the values of your aces from 1 to 14.

As long as you haven't shuffled your deck, you know just where the aces are. They appear every 13 cards. Hence, you can describe them with R's notation system:

```
deck2[c(13, 26, 39, 52), ]
##     face     suit value
##      ace   spades     1
##      ace    clubs     1
##      ace diamonds     1
##      ace   hearts     1
```

You can single out just the *values* of the aces by subsetting the columns dimension of deck2. Or, even better, you can subset the column vector deck2$value:

```
deck2[c(13, 26, 39, 52), 3]
## 1 1 1 1

deck2$value[c(13, 26, 39, 52)]
## 1 1 1 1
```

Now all you have to do is assign a new set of values to these old values. The set of new values will have to be the same size as the set of values that you are replacing. So you could save c(14, 14, 14, 14) into the ace values, or you could just save **14** and rely on R's recycling rules to expand 14 to c(14, 14, 14, 14):

```
deck2$value[c(13, 26, 39, 52)] <- c(14, 14, 14, 14)

# or

deck2$value[c(13, 26, 39, 52)] <- 14
```

Notice that the values change *in place*. You don't end up with a modified *copy* of deck2; the new values will appear inside deck2:

```
head(deck2, 13)
##     face   suit value
##     king spades    13
##    queen spades    12
##     jack spades    11
##      ten spades    10
##     nine spades     9
##    eight spades     8
##    seven spades     7
##      six spades     6
##     five spades     5
##     four spades     4
##    three spades     3
##      two spades     2
##      ace spades    14
```

The same technique will work whether you store your data in a vector, matrix, array, list, or data frame. Just describe the values that you want to change with R's notation system, then assign over those values with R's assignment operator.

Things worked very easily in this example because you knew exactly where each ace was. The cards were sorted in an orderly manner and an ace appeared every 13 rows.

But what if the deck had been shuffled? You could look through all the cards and note the locations of the aces, but that would be tedious. If your data frame were larger, it might be impossible:

```
deck3 <- shuffle(deck)
```

Where are the aces now?

```
head(deck3)
##   face     suit value
## queen    clubs    12
##  king    clubs    13
##   ace   spades     1   # an ace
##  nine    clubs     9
## seven   spades     7
## queen diamonds    12
```

Why not ask R to find the aces for you? You can do this with logical subsetting. Logical subsetting provides a way to do targeted extraction and modification with R objects, a sort of search-and-destroy mission inside your own data sets.

Logical Subsetting

Do you remember R's logical index system, "Logical Values" on page 69? To recap, you can select values with a vector of TRUEs and FALSEs. The vector must be the same length as the dimension that you wish to subset. R will return every element that matches a TRUE:

```
vec
## 1 0 1 1 2 1 0

vec[c(FALSE, FALSE, FALSE, FALSE, TRUE, FALSE, FALSE)]
## 2
```

At first glance, this system might seem impractical. Who wants to type out long vectors of TRUEs and FALSEs? No one. But you don't have to. You can let a logical test create a vector of TRUEs and FALSEs for you.

Logical Tests

A logical test is a comparison like "is one less than two?", 1 < 2, or "is three greater than four?", 3 > 4. R provides seven logical operators that you can use to make comparisons, shown in Table 5-1.

Table 5-1. R's logical operators

Operator	Syntax	Tests
>	a > b	Is a greater than b?
>=	a >= b	Is a greater than or equal to b?
<	a < b	Is a less than b?
<=	a <= b	Is a less than or equal to b?
==	a == b	Is a equal to b?
!=	a != b	Is a not equal to b?
%in%	a %in% c(a, b, c)	Is a in the group c(a, b, c)?

Each operator returns a TRUE or a FALSE. If you use an operator to compare vectors, R will do element-wise comparisons—just like it does with the arithmetic operators:

```
1 > 2
## FALSE

1 > c(0, 1, 2)
## TRUE FALSE FALSE

c(1, 2, 3) == c(3, 2, 1)
## FALSE TRUE FALSE
```

%in% is the only operator that does not do normal element-wise execution. %in% tests whether the value(s) on the left side are in the vector on the right side. If you provide a vector on the left side, %in% will *not* pair up the values on the left with the values on the right and then do element-wise tests. Instead, %in% will independently test whether each value on the left is *somewhere* in the vector on the right:

```
1 %in% c(3, 4, 5)
## FALSE

c(1, 2) %in% c(3, 4, 5)
## FALSE FALSE

c(1, 2, 3) %in% c(3, 4, 5)
## FALSE FALSE  TRUE

c(1, 2, 3, 4) %in% c(3, 4, 5)
## FALSE FALSE  TRUE  TRUE
```

Notice that you test for equality with a double equals sign, ==, and not a single equals sign, =, which is another way to write <-. It is easy to forget and use a = b to test if a equals b. Unfortunately, you'll be in for a nasty surprise. R won't return a TRUE or FALSE, because it won't have to: a *will* equal b, because you just ran the equivalent of a <- b.

= is an assignment operator

Be careful not to confuse = with ==. = does the same thing as <-: it assigns a value to an object.

You can compare any two R objects with a logical operator; however, logical operators make the most sense if you compare two objects of the same data type. If you compare objects of different data types, R will use its coercion rules to coerce the objects to the same type before it makes the comparison.

> **Exercise**
>
> Extract the `face` column of `deck2` and test whether each value is equal to `ace`. As a challenge, use R to quickly count how many cards are equal to `ace`.

You can extract the `face` column with R's `$` notation:

```
deck2$face
##  "king"  "queen" "jack"  "ten"   "nine"
##  "eight" "seven" "six"   "five"  "four"
##  "three" "two"   "ace"   "king"  "queen"
##  "jack"  "ten"   "nine"  "eight" "seven"
##  "six"   "five"  "four"  "three" "two"
##  "ace"   "king"  "queen" "jack"  "ten"
##  "nine"  "eight" "seven" "six"   "five"
##  "four"  "three" "two"   "ace"   "king"
##  "queen" "jack"  "ten"   "nine"  "eight"
##  "seven" "six"   "five"  "four"  "three"
##  "two"   "ace"
```

Next, you can use the == operator to test whether each value is equal to `ace`. In the following code, R will use its recycling rules to individually compare every value of `deck2$face` to "ace". Notice that the quotation marks are important. If you leave them out, R will try to find an object named `ace` to compare against `deck2$face`:

```
deck2$face == "ace"
##  FALSE FALSE FALSE FALSE FALSE FALSE FALSE
##  FALSE FALSE FALSE FALSE FALSE  TRUE FALSE
##  FALSE FALSE FALSE FALSE FALSE FALSE FALSE
##  FALSE FALSE FALSE FALSE  TRUE FALSE FALSE
##  FALSE FALSE FALSE FALSE FALSE FALSE FALSE
##  FALSE FALSE FALSE  TRUE FALSE FALSE FALSE
##  FALSE FALSE FALSE FALSE FALSE FALSE FALSE
##  FALSE FALSE  TRUE
```

You can use sum to quickly count the number of TRUEs in the previous vector. Remember that R will coerce logicals to numerics when you do math with them. R will turn TRUEs into ones and FALSEs into zeroes. As a result, sum will count the number of TRUEs:

```
sum(deck2$face == "ace")
## 4
```

You can use this method to spot and then change the aces in your deck—even if you've shuffled your cards. First, build a logical test that identifies the aces in your shuffled deck:

```
deck3$face == "ace"
```

Then use the test to single out the ace point values. Since the test returns a logical vector, you can use it as an index:

```
deck3$value[deck3$face == "ace"]
##   1 1 1 1
```

Finally, use assignment to change the ace values in deck3:

```
deck3$value[deck3$face == "ace"] <- 14

head(deck3)
##    face     suit value
## queen    clubs    12
##  king    clubs    13
##   ace   spades    14  # an ace
##  nine    clubs     9
## seven   spades     7
## queen diamonds    12
```

To summarize, you can use a logical test to select values within an object.

Logical subsetting is a powerful technique because it lets you quickly identify, extract, and modify individual values in your data set. When you work with logical subsetting, you do not need to know *where* in your data set a value exists. You only need to know how to describe the value with a logical test.

Logical subsetting is one of the things R does best. In fact, logical subsetting is a key component of vectorized programming, a coding style that lets you write fast and efficient R code, which we will study in Chapter 10.

Let's put logical subsetting to use with a new game: hearts. In hearts, every card has a value of zero:

```
deck4 <- deck
deck4$value <- 0

head(deck4, 13)
##   face  suit value
##   king spades     0
##  queen spades     0
```

```
##    jack spades     0
##     ten spades     0
##    nine spades     0
##   eight spades     0
##   seven spades     0
##     six spades     0
##    five spades     0
##    four spades     0
##   three spades     0
##     two spades     0
##     ace spades     0
```

except cards in the suit of hearts and the queen of spades. Each card in the suit of hearts has a value of 1. Can you find these cards and replace their values? Give it a try.

> **Exercise**
>
> Assign a value of 1 to every card in deck4 that has a suit of hearts.

To do this, first write a test that identifies cards in the hearts suit:

```
deck4$suit == "hearts"
##  FALSE FALSE FALSE FALSE FALSE FALSE FALSE
##  FALSE FALSE FALSE FALSE FALSE FALSE FALSE
##  FALSE FALSE FALSE FALSE FALSE FALSE FALSE
##  FALSE FALSE FALSE FALSE FALSE FALSE FALSE
##  FALSE FALSE FALSE FALSE FALSE FALSE FALSE
##  FALSE FALSE FALSE FALSE  TRUE  TRUE  TRUE
##   TRUE  TRUE  TRUE  TRUE  TRUE  TRUE  TRUE
##   TRUE  TRUE  TRUE
```

Then use your test to select the values of these cards:

```
deck4$value[deck4$suit == "hearts"]
## 0 0 0 0 0 0 0 0 0 0 0 0 0
```

Finally, assign a new number to these values:

```
deck4$value[deck4$suit == "hearts"] <- 1
```

Now all of your hearts cards have been updated:

```
deck4$value[deck4$suit == "hearts"]
## 1 1 1 1 1 1 1 1 1 1 1 1 1
```

In hearts, the queen of spades has the most unusual value of all: she's worth 13 points. It should be simple to change her value, but she's surprisingly hard to find. You could find all of the *queens*:

```
deck4[deck4$face == "queen", ]
##    face   suit value
##   queen spades     0
```

```
## queen    clubs    0
## queen diamonds    0
## queen   hearts    1
```

But that's three cards too many. On the other hand, you could find all of the cards in *spades*:

```
deck4[deck4$suit == "spades", ]
##   face  suit value
##   king spades    0
##  queen spades    0
##   jack spades    0
##    ten spades    0
##   nine spades    0
##  eight spades    0
##  seven spades    0
##    six spades    0
##   five spades    0
##   four spades    0
##  three spades    0
##    two spades    0
##    ace spades    0
```

But that's 12 cards too many. What you really want to find is all of the cards that have both a face value equal to queen and a suit value equal to spades. You can do that with a *Boolean operator*. Boolean operators combine multiple logical tests together into a single test.

Boolean Operators

Boolean operators are things like *and* (&) and *or* (|). They collapse the results of multiple logical tests into a single TRUE or FALSE. R has six boolean operators, shown in Table 5-2.

Table 5-2. R's Boolean operators

Operator	Syntax	Tests
&	cond1 & cond2	Are both cond1 and cond2 true?
|	cond1 pipe cond2	Is one or more of cond1 and cond2 true?
xor	xor(cond1, cond2)	Is exactly one of cond1 and cond2 true?
!	!cond1	Is cond1 false? (e.g., ! flips the results of a logical test)
any	any(cond1, cond2, cond3, ...)	Are any of the conditions true?
all	all(cond1, cond2, cond3, ...)	Are all of the conditions true?

To use a Boolean operator, place it between two *complete* logical tests. R will execute each logical test and then use the Boolean operator to combine the results into a single TRUE or FALSE, Figure 5-1.

The most common mistake with Boolean operators
It is easy to forget to put a complete test on either side of a Boolean operator. In English, it is efficient to say "Is *x* greater than two and less than nine?" But in R, you need to write the equivalent of "Is *x* greater than two and *is x* less than nine?" This is shown in Figure 5-1.

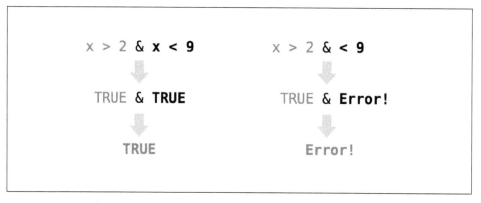

Figure 5-1. R will evaluate the expressions on each side of a Boolean operator separately, and then combine the results into a single TRUE or FALSE. If you do not supply a complete test to each side of the operator, R will return an error.

When used with vectors, Boolean operators will follow the same element-wise execution as arithmetic and logical operators:

```
a <- c(1, 2, 3)
b <- c(1, 2, 3)
c <- c(1, 2, 4)

a == b
##   TRUE TRUE TRUE

b == c
##   TRUE  TRUE FALSE

a == b & b == c
##   TRUE  TRUE FALSE
```

Could you use a Boolean operator to locate the queen of spades in your deck? Of course you can. You want to test each card to see if it is both a queen *and* a spade. You can write this test in R with:

```
deck4$face == "queen" & deck4$suit == "spades"
##   FALSE  TRUE FALSE FALSE FALSE FALSE FALSE
##   FALSE FALSE FALSE FALSE FALSE FALSE FALSE
##   FALSE FALSE FALSE FALSE FALSE FALSE FALSE
```

```
##    FALSE FALSE FALSE FALSE FALSE FALSE FALSE
##    FALSE FALSE FALSE FALSE FALSE FALSE FALSE
##    FALSE FALSE FALSE FALSE FALSE FALSE FALSE
##    FALSE FALSE FALSE FALSE FALSE FALSE FALSE
##    FALSE FALSE FALSE
```

I'll save the results of this test to its own object. That will make the results easier to work with:

```
queenOfSpades <- deck4$face == "queen" & deck4$suit == "spades"
```

Next, you can use the test as an index to select the value of the queen of spades. Make sure the test actually selects the correct value:

```
deck4[queenOfSpades, ]
##        face   suit value
## queen spades     0

deck4$value[queenOfSpades]
## 0
```

Now that you've found the queen of spades, you can update her value:

```
deck4$value[queenOfSpades] <- 13

deck4[queenOfSpades, ]
##        face   suit value
## queen spades    13
```

Your deck is now ready to play hearts.

Exercise

If you think you have the hang of logical tests, try converting these sentences into tests written with R code. To help you out, I've defined some R objects after the sentences that you can use these to test your answers:

- Is w positive?
- Is x greater than 10 and less than 20?
- Is object y the word February?
- Is *every* value in z a day of the week?

```
w <- c(-1, 0, 1)
x <- c(5, 15)
y <- "February"
z <- c("Monday", "Tuesday", "Friday")
```

Here are some model answers. If you got stuck, be sure to re-read how R evaluates logical tests that use Boolean values:

```
w > 0
10 < x & x < 20
y == "February"
all(z %in% c("Monday", "Tuesday", "Wednesday", "Thursday", "Friday",
  "Saturday", "Sunday"))
```

Let's consider one last game, blackjack. In blackjack, each number card has a value equal to its face value. Each face card (king, queen, or jack) has a value of 10. Finally, each ace has a value of 11 or 1, depending on the final results of the game.

Let's begin with a fresh copy of deck—that way the number cards (two through ten) will start off with the correct value:

```
deck5 <- deck

head(deck5, 13)
##    king spades    13
##   queen spades    12
##    jack spades    11
##     ten spades    10
##    nine spades     9
##   eight spades     8
##   seven spades     7
##     six spades     6
##    five spades     5
##    four spades     4
##   three spades     3
##     two spades     2
##     ace spades     1
```

You can change the value of the face cards in one fell swoop with %in%:

```
facecard <- deck5$face %in% c("king", "queen", "jack")

deck5[facecard, ]
##    face     suit value
##    king   spades    13
##   queen   spades    12
##    jack   spades    11
##    king    clubs    13
##   queen    clubs    12
##    jack    clubs    11
##    king diamonds    13
##   queen diamonds    12
##    jack diamonds    11
##    king   hearts    13
##   queen   hearts    12
##    jack   hearts    11

deck5$value[facecard] <- 10

head(deck5, 13)
##    face     suit value
```

```
##   king spades   10
##  queen spades   10
##   jack spades   10
##    ten spades   10
##   nine spades    9
##  eight spades    8
##  seven spades    7
##    six spades    6
##   five spades    5
##   four spades    4
##  three spades    3
##    two spades    2
##    ace spades    1
```

Now you just need to fix the ace values—or do you? It is hard to decide what value to give the aces because their exact value will change from hand to hand. At the end of each hand, an ace will equal 11 if the sum of the player's cards does not exceed 21. Otherwise, the ace will equal 1. The actual value of the ace will depend on the other cards in the player's hand. This is a case of missing information. At the moment, you do not have enough information to assign a correct point value to the ace cards.

Missing Information

Missing information problems happen frequently in data science. Usually, they are more straightforward: you don't know a value because the measurement was lost, corrupted, or never taken to begin with. R has a way to help you manage these missing values.

The NA character is a special symbol in R. It stands for "not available" and can be used as a placeholder for missing information. R will treat NA exactly as you should want missing information treated. For example, what result would you expect if you add 1 to a piece of missing information?

```
1 + NA
## NA
```

R will return a second piece of missing information. It would not be correct to say that 1 + NA = 1 because there is a good chance that the missing quantity is not zero. You do not have enough information to determine the result.

What if you tested whether a piece of missing information is equal to 1?

```
NA == 1
## NA
```

Again, your answer would be something like "I do not know if this is equal to one," that is, NA. Generally, NAs will propagate whenever you use them in an R operation or function. This can save you from making errors based on missing data.

na.rm

Missing values can help you work around holes in your data sets, but they can also create some frustrating problems. Suppose, for example, that you've collected 1,000 observations and wish to take their average with R's mean function. If even one of the values is NA, your result will be NA:

```
c(NA, 1:50)
## NA  1  2  3  4  5  6  7  8  9 10 11 12 13 14 15 16
## 17 18 19 20 21 22 23 24 25 26 27 28 29 30 31 32 33
## 34 35 36 37 38 39 40 41 42 43 44 45 46 47 48 49 50

mean(c(NA, 1:50))
## NA
```

Understandably, you may prefer a different behavior. Most R functions come with the optional argument, na.rm, which stands for NA remove. R will ignore NAs when it evaluates a function if you add the argument na.rm = TRUE:

```
mean(c(NA, 1:50), na.rm = TRUE)
## 25.5
```

is.na

On occasion, you may want to identify the NAs in your data set with a logical test, but that too creates a problem. How would you go about it? If something is a missing value, any logical test that uses it will return a missing value, even this test:

```
NA == NA
## NA
```

Which means that tests like this won't help you find missing values:

```
c(1, 2, 3, NA) == NA
## NA NA NA NA
```

But don't worry too hard; R supplies a special function that can test whether a value is an NA. The function is sensibly named is.na:

```
is.na(NA)
## TRUE

vec <- c(1, 2, 3, NA)
is.na(vec)
## FALSE FALSE FALSE  TRUE
```

Let's set all of your ace values to NA. This will accomplish two things. First, it will remind you that you do not know the final value of each ace. Second, it will prevent you from accidentally scoring a hand that has an ace before you determine the ace's final value.

You can set your ace values to NA in the same way that you would set them to a number:

```
deck5$value[deck5$face == "ace"] <- NA

head(deck5, 13)
##    face   suit value
##    king spades    10
##   queen spades    10
##    jack spades    10
##     ten spades    10
##    nine spades     9
##   eight spades     8
##   seven spades     7
##     six spades     6
##    five spades     5
##    four spades     4
##   three spades     3
##     two spades     2
##     ace spades    NA
```

Congratulations. Your deck is now ready for a game of blackjack.

Summary

You can modify values in place inside an R object when you combine R's notation syntax with the assignment operator, <-. This lets you update your data and clean your data sets.

When you work with large data sets, modifying and retrieving values creates a logistical problem of its own. How can you search through the data to find the values that you want to modify or retrieve? As an R user, you can do this with logical subsetting. Create a logical test with logical and Boolean operators and then use the test as an index in R's bracket notation. R will return the values that you are looking for, even if you do not know where they are.

Retrieving individual values will not be your only concern as an R programmer. You'll also need to retrieve entire data sets themselves; for example, you may call one in a function. Chapter 6 will teach you how R looks up and saves data sets and other R objects in its environment system. You'll then use this knowledge to fix the deal and shuffle functions.

CHAPTER 6
Environments

Your deck is now ready for a game of blackjack (or hearts or war), but are your `shuffle` and `deal` functions up to snuff? Definitely not. For example, `deal` deals the same card over and over again:

```
deal(deck)
## face    suit value
## king spades    13

deal(deck)
## face    suit value
## king spades    13

deal(deck)
## face    suit value
## king spades    13
```

And the `shuffle` function doesn't actually shuffle `deck` (it returns a copy of `deck` that has been shuffled). In short, both of these functions use `deck`, but neither manipulates `deck`—and we would like them to.

To fix these functions, you will need to learn how R stores, looks up, and manipulates objects like `deck`. R does all of these things with the help of an environment system.

Environments

Consider for a moment how your computer stores files. Every file is saved in a folder, and each folder is saved in another folder, which forms a hierarchical file system. If your computer wants to open up a file, it must first look up the file in this file system.

You can see your file system by opening a finder window. For example, Figure 6-1 shows part of the file system on my computer. I have tons of folders. Inside one of them is a subfolder named *Documents*, inside of that subfolder is a sub-subfolder named *ggsubplot*, inside of that folder is a folder named *inst*, inside of that is a folder named *doc*, and inside of that is a file named *manual.pdf*.

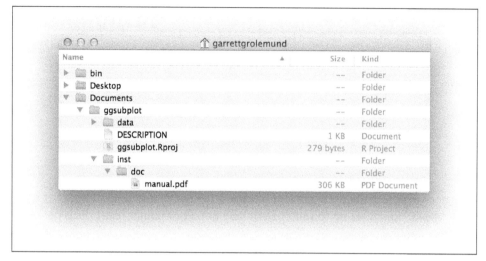

Figure 6-1. Your computer arranges files into a hierarchy of folders and subfolders. To look at a file, you need to find where it is saved in the file system.

R uses a similar system to save R objects. Each object is saved inside of an environment, a list-like object that resembles a folder on your computer. Each environment is connected to a *parent environment,* a higher-level environment, which creates a hierarchy of environments.

You can see R's environment system with the `parenvs` function in the pryr package. `parenvs(all = TRUE)` will return a list of the environments that your R session is using. The actual output will vary from session to session depending on which packages you have loaded. Here's the output from my current session:

```
library(pryr)
parenvs(all = TRUE)
##    label                                name
## 1  <environment: R_GlobalEnv>           ""
## 2  <environment: package:pryr>          "package:pryr"
## 3  <environment: 0x7fff3321c388>        "tools:rstudio"
## 4  <environment: package:stats>         "package:stats"
## 5  <environment: package:graphics>      "package:graphics"
## 6  <environment: package:grDevices>     "package:grDevices"
## 7  <environment: package:utils>         "package:utils"
## 8  <environment: package:datasets>      "package:datasets"
## 9  <environment: package:methods>       "package:methods"
## 10 <environment: 0x7fff3193dab0>        "Autoloads"
## 11 <environment: base>                  ""
## 12 <environment: R_EmptyEnv>            ""
```

It takes some imagination to interpret this output, so let's visualize the environments as a system of folders, Figure 6-2. You can think of the environment tree like this. The lowest-level environment is named R_GlobalEnv and is saved inside an environment named package:pryr, which is saved inside the environment named 0x7fff3321c388, and so on, until you get to the final, highest-level environment, R_EmptyEnv. R_EmptyEnv is the only R environment that does not have a parent environment.

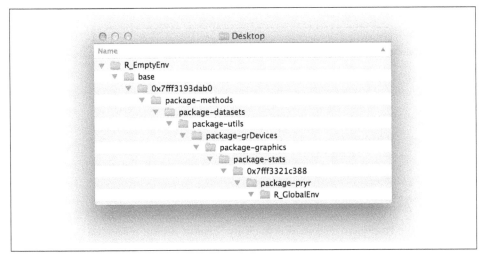

Figure 6-2. R stores R objects in an environment tree that resembles your computer's folder system.

Remember that this example is just a metaphor. R's environments exist in your RAM memory, and not in your file system. Also, R environments aren't technically saved inside one another. Each environment is connected to a parent environment, which makes it easy to search up R's environment tree. But this connection is one-way: there's no way to look at one environment and tell what its "children" are. So you cannot search down R's environment tree. In other ways, though, R's environment system works similar to a file system.

Working with Environments

R comes with some helper functions that you can use to explore your environment tree. First, you can refer to any of the environments in your tree with `as.environment`. `as.environment` takes an environment name (as a character string) and returns the corresponding environment:

```
as.environment("package:stats")
## <environment: package:stats>
## attr(,"name")
```

```
## [1] "package:stats"
## attr(,"path")
## [1] "/Library/Frameworks/R.framework/Versions/3.0/Resources/library/stats"
```

Three environments in your tree also come with their own accessor functions. These are the global environment (R_GlobalEnv), the base environment (base), and the empty environment (R_EmptyEnv). You can refer to them with:

```
globalenv()
## <environment: R_GlobalEnv>

baseenv()
## <environment: base>

emptyenv()
##<environment: R_EmptyEnv>
```

Next, you can look up an environment's parent with parent.env:

```
parent.env(globalenv())
## <environment: package:pryr>
## attr(,"name")
## [1] "package:pryr"
## attr(,"path")
## [1] "/Library/Frameworks/R.framework/Versions/3.0/Resources/library/pryr"
```

Notice that the empty environment is the only R environment without a parent:

```
parent.env(emptyenv())
## Error in parent.env(emptyenv()) : the empty environment has no parent
```

You can view the objects saved in an environment with ls or ls.str. ls will return just the object names, but ls.str will display a little about each object's structure:

```
ls(emptyenv())
## character(0)

ls(globalenv())
##  "deal"    "deck"     "deck2"    "deck3"   "deck4"  "deck5"
##  "die"     "gender"   "hand"     "lst"     "mat"    "mil"
##  "new"     "now"      "shuffle"  "vec"
```

The empty environment is—not surprisingly—empty; the base environment has too many objects to list here; and the global environment has some familiar faces. It is where R has saved all of the objects that you've created so far.

RStudio's environment pane displays all of the objects in your global environment.

You can use R's $ syntax to access an object in a specific environment. For example, you can access deck from the global environment:

```
head(globalenv()$deck, 3)
##    face  suit value
## king    spades    13
## queen   spades    12
## jack    spades    11
```

And you can use the `assign` function to save an object into a particular environment. First give `assign` the name of the new object (as a character string). Then give `assign` the value of the new object, and finally the environment to save the object in:

```
assign("new", "Hello Global", envir = globalenv())

globalenv()$new
## "Hello Global"
```

Notice that `assign` works similar to <-. If an object already exists with the given name in the given environment, `assign` will overwrite it without asking for permission. This makes `assign` useful for updating objects but creates the potential for heartache.

Now that you can explore R's environment tree, let's examine how R uses it. R works closely with the environment tree to look up objects, store objects, and evaluate functions. How R does each of these tasks will depend on the current active environment.

The Active Environment

At any moment of time, R is working closely with a single environment. R will store new objects in this environment (if you create any), and R will use this environment as a starting point to look up existing objects (if you call any). I'll call this special environment the *active environment*. The active environment is usually the global environment, but this may change when you run a function.

You can use `environment` to see the current active environment:

```
environment()
<environment: R_GlobalEnv>
```

The global environment plays a special role in R. It is the active environment for every command that you run at the command line. As a result, any object that you create at the command line will be saved in the global environment. You can think of the global environment as your user workspace.

When you call an object at the command line, R will look for it first in the global environment. But what if the object is not there? In that case, R will follow a series of rules to look up the object.

Scoping Rules

R follows a special set of rules to look up objects. These rules are known as R's scoping rules, and you've already met a couple of them:

1. R looks for objects in the current active environment.
2. When you work at the command line, the active environment is the global environment. Hence, R looks up objects that you call at the command line in the global environment.

Here is a third rule that explains how R finds objects that are not in the active environment:

3. When R does not find an object in an environment, R looks in the environment's parent environment, then the parent of the parent, and so on, until R finds the object or reaches the empty environment.

So, if you call an object at the command line, R will look for it in the global environment. If R can't find it there, R will look in the parent of the global environment, and then the parent of the parent, and so on, working its way up the environment tree until it finds the object, as in Figure 6-3. If R cannot find the object in any environment, it will return an error that says the object is not found.

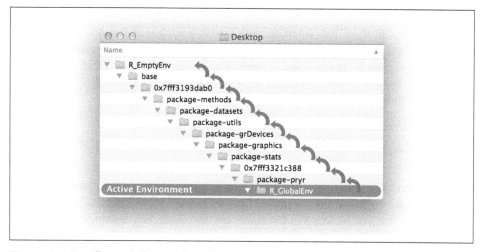

Figure 6-3. R will search for an object by name in the active environment, here the global environment. If R does not find the object there, it will search in the active environment's parent, and then the parent's parent, and so on until R finds the object or runs out of environments.

Remember that functions are a type of object in R. R will store and look up functions the same way it stores and looks up other objects, by searching for them by name in the environment tree.

Assignment

When you assign a value to an object, R saves the value in the active environment under the object's name. If an object with the same name already exists in the active environment, R will overwrite it.

For example, an object named new exists in the global environment:

```
new
## "Hello Global"
```

You can save a new object named new to the global environment with this command. R will overwrite the old object as a result:

```
new <- "Hello Active"
```

```
new
## "Hello Active"
```

This arrangement creates a quandry for R whenever R runs a function. Many functions save temporary objects that help them do their jobs. For example, the roll function from Part I saved an object named die and an object named dice:

```
roll <- function() {
  die <- 1:6
  dice <- sample(die, size = 2, replace = TRUE)
  sum(dice)
}
```

R must save these temporary objects in the active environment; but if R does that, it may overwrite existing objects. Function authors cannot guess ahead of time which names may already exist in your active environment. How does R avoid this risk? Every time R runs a function, it creates a new active environment to evaluate the function in.

Evaluation

R creates a new environment *each* time it evaluates a function. R will use the new environment as the active environment while it runs the function, and then R will return to the environment that you called the function from, bringing the function's result with it. Let's call these new environments *runtime environments* because R creates them at runtime to evaluate functions.

We'll use the following function to explore R's runtime environments. We want to know what the environments look like: what are their parent environments, and what objects do they contain? show_env is designed to tell us:

```
show_env <- function(){
  list(ran.in = environment(),
    parent = parent.env(environment()),
    objects = ls.str(environment()))
}
```

show_env is itself a function, so when we call show_env(), R will create a runtime environment to evaluate the function in. The results of show_env will tell us the name of the runtime environment, its parent, and which objects the runtime environment contains:

```
show_env()
## $ran.in
## <environment: 0x7ff711d12e28>
## 
## $parent
## <environment: R_GlobalEnv>
## 
## $objects
```

The results reveal that R created a new environment named 0x7ff711d12e28 to run show_env() in. The environment had no objects in it, and its parent was the global environment. So for purposes of running show_env, R's environment tree looked like Figure 6-4.

Let's run show_env again:

```
show_env()
## $ran.in
## <environment: 0x7ff715f49808>
## 
## $parent
## <environment: R_GlobalEnv>
## 
## $objects
```

This time show_env ran in a new environment, 0x7ff715f49808. R creates a new environment *each* time you run a function. The 0x7ff715f49808 environment looks exactly the same as 0x7ff711d12e28. It is empty and has the same global environment as its parent.

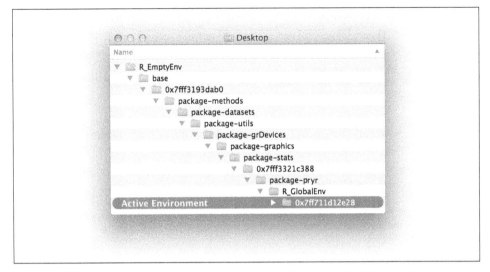

Figure 6-4. R creates a new environment to run show_env in. The environment is a child of the global environment.

Now let's consider which environment R will use as the parent of the runtime environment.

R will connect a function's runtime environment to the environment that the function *was first created in*. This environment plays an important role in the function's life—because all of the function's runtime environments will use it as a parent. Let's call this environment the *origin environment*. You can look up a function's origin environment by running `environment` on the function:

```
environment(show_env)
## <environment: R_GlobalEnv>
```

The origin environment of `show_env` is the global environment because we created `show_env` at the command line, but the origin environment does not need to be the global environment. For example, the environment of `parenvs` is the `pryr` package:

```
environment(parenvs)
## <environment: namespace:pryr>
```

In other words, the parent of a runtime environment will not always be the global environment; it will be whichever environment the function was first created in.

Finally, let's look at the objects contained in a runtime environment. At the moment, `show_env`'s runtime environments do not contain any objects, but that is easy to fix. Just have `show_env` create some objects in its body of code. R will store any objects created by `show_env` in its runtime environment. Why? Because the runtime environment will be the active environment when those objects are created:

Evaluation | 101

```
show_env <- function(){
  a <- 1
  b <- 2
  c <- 3
  list(ran.in = environment(),
    parent = parent.env(environment()),
    objects = ls.str(environment()))
}
```

This time when we run show_env, R stores a, b, and c in the runtime environment:

```
show_env()
## $ran.in
## <environment: 0x7ff712312cd0>
##
## $parent
## <environment: R_GlobalEnv>
##
## $objects
## a :  num 1
## b :  num 2
## c :  num 3
```

This is how R ensures that a function does not overwrite anything that it shouldn't. Any objects created by the function are stored in a safe, out-of-the-way runtime environment.

R will also put a second type of object in a runtime environment. If a function has arguments, R will copy over each argument to the runtime environment. The argument will appear as an object that has the name of the argument but the value of whatever input the user provided for the argument. This ensures that a function will be able to find and use each of its arguments:

```
foo <- "take me to your runtime"

show_env <- function(x = foo){
  list(ran.in = environment(),
    parent = parent.env(environment()),
    objects = ls.str(environment()))
}

show_env()
## $ran.in
## <environment: 0x7ff712398958>
##
## $parent
## <environment: R_GlobalEnv>
##
## $objects
## x :  chr "take me to your runtime"
```

Let's put this all together to see how R evaluates a function. Before you call a function, R is working in an active environment; let's call this the *calling environment*. It is the environment R calls the function from.

Then you call the function. R responds by setting up a new runtime environment. This environment will be a child of the function's origin enviornment. R will copy each of the function's arguments into the runtime environment and then make the runtime environment the new active environment.

Next, R runs the code in the body of the function. If the code creates any objects, R stores them in the active, that is, runtime environment. If the code calls any objects, R uses its scoping rules to look them up. R will search the runtime environment, then the parent of the runtime environment (which will be the origin environment), then the parent of the origin environment, and so on. Notice that the calling environment might not be on the search path. Usually, a function will only call its arguments, which R can find in the active runtime environment.

Finally, R finishes running the function. It switches the active environment back to the calling environment. Now R executes any other commands in the line of code that called the function. So if you save the result of the function to an object with <-, the new object will be stored in the calling environment.

To recap, R stores its objects in an environment system. At any moment of time, R is working closely with a single active environment. It stores new objects in this environment, and it uses the environment as a starting point when it searches for existing objects. R's active environment is usually the global environment, but R will adjust the active environment to do things like run functions in a safe manner.

How can you use this knowledge to fix the `deal` and `shuffle` functions?

First, let's start with a warm-up question. Suppose I redefine `deal` at the command line like this:

```
deal <- function() {
  deck[1, ]
}
```

Notice that `deal` no longer takes an argument, and it calls the `deck` object, which lives in the global environment.

> **Exercise**
>
> Will R be able to find deck and return an answer when I call the new version of `deal`, such as `deal()`?

Yes. deal will still work the same as before. R will run deal in a runtime environment that is a child of the global environment. Why will it be a child of the global environment? Because the global environment is the origin environment of deal (we defined deal in the global environment):

```
environment(deal)
## <environment: R_GlobalEnv>
```

When deal calls deck, R will need to look up the deck object. R's scoping rules will lead it to the version of deck in the global environment, as in Figure 6-5. deal works as expected as a result:

```
deal()
##  face  suit   value
##  king  spades 13
```

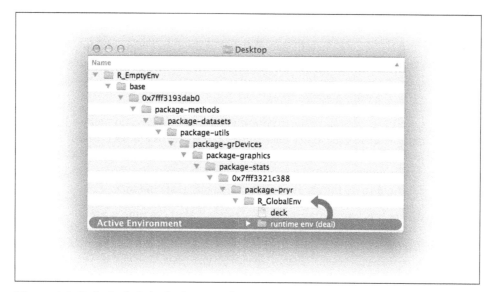

Figure 6-5. R finds deck by looking in the parent of deal's runtime environment. The parent is the global environment, deal's origin environment. Here, R finds the copy of deck.

Now let's fix the deal function to remove the cards it has dealt from deck. Recall that deal returns the top card of deck but does not remove the card from the deck. As a result, deal always returns the same card:

```
deal()
## face  suit value
## king spades   13

deal()
## face  suit value
## king spades   13
```

You know enough R syntax to remove the top card of deck. The following code will save a pristine copy of deck and then remove the top card:

```
DECK <- deck

deck <- deck[-1, ]

head(deck, 3)
## face  suit value
## queen spades  12
## jack  spades  11
## ten   spades  10
```

Now let's add the code to deal. Here deal saves (and then returns) the top card of deck. In between, it removes the card from deck…or does it?

```
deal <- function() {
  card <- deck[1, ]
  deck <- deck[-1, ]
  card
}
```

This code won't work because R will be in a runtime environment when it executes deck <- deck[-1,]. Instead of overwriting the global copy of deck with deck[-1,], deal will just create a slightly altered copy of deck in its runtime environment, as in Figure 6-6.

Exercise

Rewrite the deck <- deck[-1,] line of deal to *assign* deck[-1,] to an object named deck in the global environment. Hint: consider the assign function.

Evaluation | 105

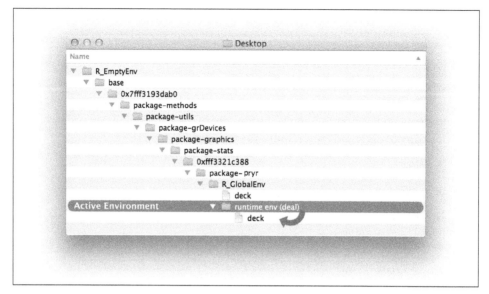

Figure 6-6. The deal function looks up deck in the global environment but saves deck[-1,] in the runtime environment as a new object named deck.

You can assign an object to a specific environment with the `assign` function:

```
deal <- function() {
  card <- deck[1, ]
  assign("deck", deck[-1, ], envir = globalenv())
  card
}
```

Now deal will finally clean up the global copy of deck, and we can deal cards just as we would in real life:

```
deal()
##  face   suit value
## queen spades    12

deal()
##  face   suit value
##  jack spades    11

deal()
##  face   suit value
##   ten spades    10
```

Let's turn our attention to the shuffle function:

```
shuffle <- function(cards) {
  random <- sample(1:52, size = 52)
  cards[random, ]
}
```

`shuffle(deck)` doesn't shuffle the `deck` object; it returns a shuffled copy of the `deck` object:

```
head(deck, 3)
##   face   suit value
## nine spades    9
## eight spades   8
## seven spades   7

a <- shuffle(deck)

head(deck, 3)
##   face   suit value
## nine spades    9
## eight spades   8
## seven spades   7

head(a, 3)
##   face    suit value
##    ace diamonds    1
## seven    clubs     7
##   two    clubs     2
```

This behavior is now undesirable in two ways. First, `shuffle` fails to shuffle deck. Second, `shuffle` returns a copy of `deck`, which may be missing the cards that have been dealt away. It would be better if `shuffle` returned the dealt cards to the deck and then shuffled. This is what happens when you shuffle a deck of cards in real life.

> **Exercise**
>
> Rewrite `shuffle` so that it replaces the copy of `deck` that lives in the global environment with a shuffled version of `DECK`, the intact copy of `deck` that also lives in the global environment. The new version of `shuffle` should have no arguments and return no output.

You can update `shuffle` in the same way that you updated `deck`. The following version will do the job:

```
shuffle <- function(){
  random <- sample(1:52, size = 52)
  assign("deck", DECK[random, ], envir = globalenv())
}
```

Since `DECK` lives in the global environment, `shuffle`'s environment of origin, `shuffle` will be able to find `DECK` at runtime. R will search for `DECK` first in `shuffle`'s runtime environment, and then in `shuffle`'s origin environment—the global environment—which is where `DECK` is stored.

The second line of shuffle will create a reordered copy of DECK and save it as deck in the global environment. This will overwrite the previous, nonshuffled version of deck.

Closures

Our system finally works. For example, you can shuffle the cards and then deal a hand of blackjack:

```
shuffle()

deal()
##   face    suit value
## queen hearts     12

deal()
##   face    suit value
## eight hearts      8
```

But the system requires deck and DECK to exist in the global environment. Lots of things happen in this environment, and it is possible that deck may get modified or erased by accident.

It would be better if we could store deck in a safe, out-of-the-way place, like one of those safe, out-of-the-way environments that R creates to run functions in. In fact, storing deck in a runtime environment is not such a bad idea.

You could create a function that takes deck as an argument and saves a copy of deck as DECK. The function could also save its own copies of deal and shuffle:

```
setup <- function(deck) {
  DECK <- deck

  DEAL <- function() {
    card <- deck[1, ]
    assign("deck", deck[-1, ], envir = globalenv())
    card
  }

  SHUFFLE <- function(){
    random <- sample(1:52, size = 52)
    assign("deck", DECK[random, ], envir = globalenv())
  }
}
```

When you run setup, R will create a runtime environment to store these objects in. The environment will look like Figure 6-7.

Now all of these things are safely out of the way in a child of the global environment. That makes them safe but hard to use. Let's ask setup to return DEAL and SHUFFLE so we can use them. The best way to do this is to return the functions as a list:

```
setup <- function(deck) {
  DECK <- deck

  DEAL <- function() {
    card <- deck[1, ]
    assign("deck", deck[-1, ], envir = globalenv())
    card
  }

  SHUFFLE <- function(){
    random <- sample(1:52, size = 52)
    assign("deck", DECK[random, ], envir = globalenv())
  }

  list(deal = DEAL, shuffle = SHUFFLE)
}

cards <- setup(deck)
```

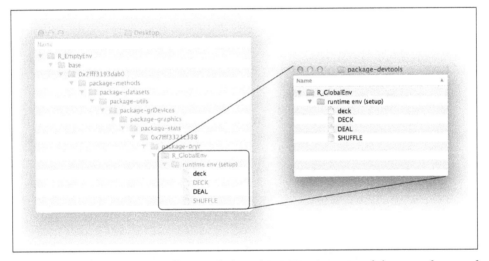

Figure 6-7. Running setup will store deck and DECK in an out-of-the-way place, and create a DEAL and SHUFFLE function. Each of these objects will be stored in an environment whose parent is the global environment.

Then you can save each of the elements of the list to a dedicated object in the global environment:

```
deal <- cards$deal
shuffle <- cards$shuffle
```

Now you can run deal and shuffle just as before. Each object contains the same code as the original deal and shuffle:

```
deal
## function() {
##     card <- deck[1, ]
##     assign("deck", deck[-1, ], envir = globalenv())
##     card
## }
## <environment: 0x7ff7169c3390>

shuffle
## function(){
##     random <- sample(1:52, size = 52)
##     assign("deck", DECK[random, ], envir = globalenv())
## }
## <environment: 0x7ff7169c3390>
```

However, the functions now have one important difference. Their origin environment is no longer the global environment (although deal and shuffle *are* currently saved there). Their origin environment is the runtime environment that R made when you ran setup. That's where R created DEAL and SHUFFLE, the functions copied into the new deal and shuffle, as shown in:

```
environment(deal)
## <environment: 0x7ff7169c3390>

environment(shuffle)
## <environment: 0x7ff7169c3390>
```

Why does this matter? Because now when you run deal or shuffle, R will evaluate the functions in a runtime environment that uses 0x7ff7169c3390 as its parent. DECK and deck will be in this parent environment, which means that deal and shuffle will be able to find them at runtime. DECK and deck will be in the functions' search path but still out of the way in every other respect, as shown in Figure 6-8.

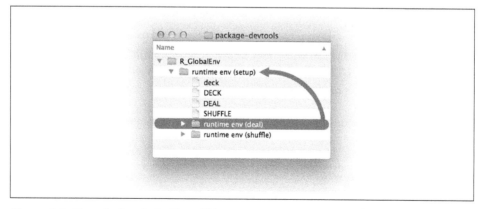

Figure 6-8. Now deal and shuffle will be run in an environment that has the protected deck and DECK in its search path.

This arrangement is called a *closure*. setup's runtime environment "encloses" the deal and shuffle functions. Both deal and shuffle can work closely with the objects contained in the enclosing environment, but almost nothing else can. The enclosing environment is not on the search path for any other R function or environment.

You may have noticed that deal and shuffle still update the deck object in the global environment. Don't worry, we're about to change that. We want deal and shuffle to work exclusively with the objects in the parent (enclosing) environment of their runtime environments. Instead of having each function reference the global environment to update deck, you can have them reference their parent environment at runtime, as shown in Figure 6-9:

```
setup <- function(deck) {
  DECK <- deck

  DEAL <- function() {
    card <- deck[1, ]
    assign("deck", deck[-1, ], envir = parent.env(environment()))
    card
  }

  SHUFFLE <- function(){
    random <- sample(1:52, size = 52)
    assign("deck", DECK[random, ], envir = parent.env(environment()))
  }

  list(deal = DEAL, shuffle = SHUFFLE)
}

cards <- setup(deck)
deal <- cards$deal
shuffle <- cards$shuffle
```

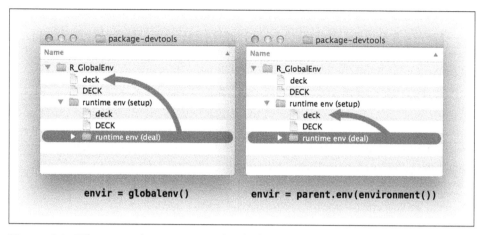

Figure 6-9. When you change your code, deal and shuffle will go from updating the global environment (left) to updating their parent environment (right).

We finally have a self-contained card game. You can delete (or modify) the global copy of deck as much as you want and still play cards. deal and shuffle will use the pristine, protected copy of deck:

```
rm(deck)

shuffle()

deal()
## face  suit value
##  ace hearts    1

deal()
## face  suit value
## jack clubs    11
```

Blackjack!

Summary

R saves its objects in an environment system that resembles your computer's file system. If you understand this system, you can predict how R will look up objects. If you call an object at the command line, R will look for the object in the global environment and then the parents of the global environment, working its way up the environment tree one environment at a time.

R will use a slightly different search path when you call an object from inside of a function. When you run a function, R creates a new environment to execute commands in. This environment will be a child of the environment where the function was originally defined. This may be the global environment, but it also may not be. You can use this behavior to create closures, which are functions linked to objects in protected environments.

As you become familiar with R's environment system, you can use it to produce elegant results, like we did here. However, the real value of understanding the environment system comes from knowing how R functions do their job. You can use this knowledge to figure out what is going wrong when a function does not perform as expected.

Project 2 Wrap-up

You now have full control over the data sets and values that you load into R. You can store data as R objects, you can retrieve and manipulate data values at will, and you can even predict how R will store and look up your objects in your computer's memory.

You may not realize it yet, but your expertise makes you a powerful, computer-augmented data user. You can use R to save and work with larger data sets than you could otherwise handle. So far we've only worked with deck, a small data set; but you can use the same techniques to work with any data set that fits in your computer's memory.

However, storing data is not the only logistical task that you will face as a data scientist. You will often want to do tasks with your data that are so complex or repetitive that they are difficult to do without a computer. Some of the things can be done with functions that already exist in R and its packages, but others cannot. You will be the most versatile as a data scientist if you can write your own programs for computers to follow. R can help you do this. When you are ready, Part III will teach you the most useful skills for writing programs in R.

PART III
Project 3: Slot Machine

Slot machines are the most popular game in modern casinos. If you've never seen one, a slot machine resembles an arcade game that has a lever on its side. For a small fee you can pull the lever, and the machine will generate a random combination of three symbols. If the correct combination appears, you can win a prize, maybe even the jackpot.

Slot machines make fantastic profits for casinos because they offer a very low payout rate. In many games, such as Blackjack and Roulette, the odds are only slightly stacked in the casino's favor. In the long run, the casino pays back 97 to 98 cents in prizes of every dollar that a gambler spends on these games. With slot machines, it is typical for a casino to only pay back 90 to 95 cents—and the casino keeps the rest. If this seems underhanded, keep in mind that slot machines are one of the most popular games at a casino; few people seem to mind. And if you consider that state lotteries have payout rates that are much closer to 50 cents on the dollar, slot machines don't look that bad.

In this project, you will build a real, working slot machine modeled after some real life Video Lottery Terminals from Manitoba, Canada. The terminals were a source of scandal in the 1990s. You'll get to the bottom of this scandal by writing a program that recreates the slot machines. You'll then do some calculations and run some simulations that reveal the true payout rate of the machines.

This project will teach you how to write programs and run simulations in R. You will also learn how to:

- Use a practical strategy to design programs
- Use `if` and `else` statements to tell R what to do when
- Create lookup tables to find values
- Use `for`, `while`, and `repeat` loops to automate repetitive operations
- Use S3 methods, R's version of Object-Oriented Programming
- Measure the speed of R code
- Write fast, vectorized R code

CHAPTER 7
Programs

In this chapter, you will build a real, working slot machine that you can play by running an R function. When you're finished, you'll be able to play it like this:

```
play()
## 0 0 DD
## $0

play()
## 7 7 7
## $80
```

The play function will need to do two things. First, it will need to randomly generate three symbols; and, second, it will need to calculate a prize based on those symbols.

The first step is easy to simulate. You can randomly generate three symbols with the sample function—just like you randomly "rolled" two dice in Part I. The following function generates three symbols from a group of common slot machine symbols: diamonds (DD), sevens (7), triple bars (BBB), double bars (BB), single bars (B), cherries (C), and zeroes (0). The symbols are selected randomly, and each symbol appears with a different probability:

```
get_symbols <- function() {
  wheel <- c("DD", "7", "BBB", "BB", "B", "C", "0")
  sample(wheel, size = 3, replace = TRUE,
    prob = c(0.03, 0.03, 0.06, 0.1, 0.25, 0.01, 0.52))
}
```

You can use get_symbols to generate the symbols used in your slot machine:

```
get_symbols()
## "BBB" "0"   "C"

get_symbols()
## "0" "0" "0"
```

117

```
get_symbols()
## "7" "0" "B"
```

`get_symbols` uses the probabilities observed in a group of video lottery terminals from Manitoba, Canada. These slot machines became briefly controversial in the 1990s, when a reporter decided to test their payout rate. The machines appeared to pay out only 40 cents on the dollar, even though the manufacturer claimed they would pay out 92 cents on the dollar. The original data collected on the machines and a description of the controversy is available online in a journal article by W. John Braun (*http://bit.ly/jse_Braun*). The controversy died down when additional testing showed that the manufacturer was correct.

The Manitoba slot machines use the complicated payout scheme shown in Table 7-1. A player will win a prize if he gets:

1. Three of the same type of symbol (except for three zeroes)
2. Three bars (of mixed variety)
3. One or more cherries

Otherwise, the player receives no prize.

The monetary value of the prize is determined by the exact combination of symbols and is further modified by the presence of diamonds. Diamonds are treated like "wild cards," which means they can be considered any other symbol if it would increase a player's prize. For example, a player who rolls 7 7 DD would earn a prize for getting three sevens. There is one exception to this rule, however: a diamond cannot be considered a cherry unless the player also gets one real cherry. This prevents a dud roll like, 0 DD 0 from being scored as 0 C 0.

Diamonds are also special in another way. Every diamond that appears in a combination doubles the amount of the final prize. So 7 7 DD would actually be scored *higher* than 7 7. Three sevens would earn you $80, but two sevens and a diamond would earn you $160. One seven and two diamonds would be even better, resulting in a prize that has been doubled twice, or $320. A jackpot occurs when a player rolls DD DD DD. Then a player earns $100 doubled three times, which is $800.

Table 7-1. Each play of the slot machine costs $1. A player's symbols determine how much they win. Diamonds (DD) are wild, and each diamond doubles the final prize. * = any symbol.

Combination	Prize($)
DD DD DD	100
7 7 7	80
BBB BBB BBB	40

Combination	Prize($)
BB BB BB	25
B B B	10
C C C	10
Any combination of bars	5
C C *	5
C * C	5
* C C	5
C * *	2
* C *	2
* * C	2

To create your play function, you will need to write a program that can take the output of get_symbols and calculate the correct prize based on Table 7-1.

In R, programs are saved either as R scripts or as functions. We'll save your program as a function named score. When you are finished, you will be able to use score to calculate a prize like this:

```
score(c("DD", "DD", "DD"))
## 800
```

After that it will be easy to create the full slot machine, like this:

```
play <- function() {
  symbols <- get_symbols()
  print(symbols) ❶
  score(symbols)
}
```

❶ The print command prints its output to the console window, which makes print a useful way to display messages from within the body of a function.

You may notice that play calls a new function, print. This will help play display the three slot machine symbols, since they do not get returned by the last line of the function. The print command prints its output to the console window — even if R calls it from within a function.

In Part I, I encouraged you to write all of your R code in an R script, a text file where you can compose and save code. That advice will become very important as you work through this chapter. Remember that you can open an R script in RStudio by going to the menu bar and clicking on File > New File > R Script.

Strategy

Scoring slot-machine results is a complex task that will require a complex algorithm. You can make this, and other coding tasks, easier by using a simple strategy:

1. Break complex tasks into simple subtasks.
2. Use concrete examples.
3. Describe your solutions in English, then convert them to R.

Let's start by looking at how you can divide a program into subtasks that are simple to work with.

A program is a set of step-by-step instructions for your computer to follow. Taken together, these instructions may accomplish something very sophisticated. Taken apart, each individual step will likely be simple and straightforward.

You can make coding easier by identifying the individual steps or subtasks within your program. You can then work on each subtask separately. If a subtask seems complicated, try to divide it again into smaller subtasks that are even more simple. You can often reduce an R program into substasks so simple that each can be performed with a pre-existing function.

R programs contain two types of subtasks: sequential steps and parallel cases.

Sequential Steps

One way to subdivide a program is into a series of sequential steps. The play function takes the approach, shown in Figure 7-1. First, it generates three symbols (step 1), then it displays them in the console window (step 2), and then it scores them (step 3):

```
play <- function() {

  # step 1: generate symbols
  symbols <- get_symbols()

  # step 2: display the symbols
  print(symbols)

  # step 3: score the symbols
  score(symbols)
}
```

To have R execute steps in sequence, place the steps one after another in an R script or function body.

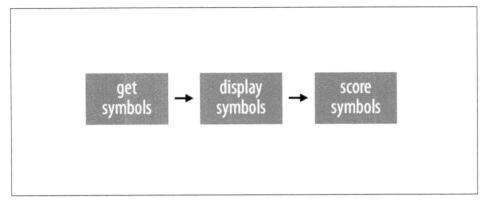

Figure 7-1. The play function uses a series of steps.

Parallel Cases

Another way to divide a task is to spot groups of similar cases within the task. Some tasks require different algorithms for different groups of input. If you can identify those groups, you can work out their algorithms one at a time.

For example, score will need to calculate the prize one way if symbols contains three of a kind (In that case, score will need to match the common symbol to a prize). score will need to calculate the prize a second way if the symbols are all bars (In that case, score can just assign a prize of $5). And, finally, score will need to calculate the prize in a third way if the symbols do not contain three of a kind or all bars (In that case, score must count the number of cherries present). score will never use all three of these algorithms at once; it will always choose just one algorithm to run based on the combination of symbols.

Diamonds complicate all of this because diamonds can be treated as wild cards. Let's ignore that for now and focus on the simpler case where diamonds double the prize but are not wilds. score can double the prize as necessary after it runs one of the following algorithms, as shown in Figure 7-2.

Adding the score cases to the play steps reveals a strategy for the complete slot machine program, as shown in Figure 7-3.

We've already solved the first few steps in this strategy. Our program can get three slot machine symbols with the get_symbols function. Then it can display the symbols with the print function. Now let's examine how the program can handle the parallel score cases.

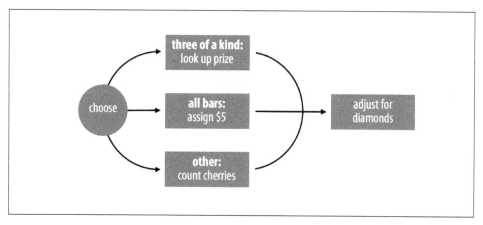

Figure 7-2. The score function must distinguish between parallel cases.

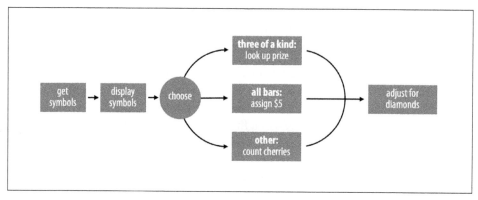

Figure 7-3. The complete slot machine simulation will involve subtasks that are arranged both in series and in parallel.

if Statements

Linking cases together in parallel requires a bit of structure; your program faces a fork in the road whenever it must choose between cases. You can help the program navigate this fork with an `if` statement.

An `if` statement tells R to do a certain task for a certain case. In English you would say something like, "If this is true, do that." In R, you would say:

```
if (this) {
  that
}
```

The this object should be a logical test or an R expression that evaluates to a single TRUE or FALSE. If this evaluates to TRUE, R will run all of the code that appears between the braces that follow the if statement (i.e., between the { and } symbols). If this evaluates to FALSE, R will skip the code between the braces without running it.

For example, you could write an if statement that ensures some object, num, is positive:

```
if (num < 0) {
  num <- num * -1
}
```

If num < 0 is TRUE, R will multiply num by negative one, which will make num positive:

```
num <- -2

if (num < 0) {
  num <- num * -1
}

num
## 2
```

If num < 0 is FALSE, R will do nothing and num will remain as it is—positive (or zero):

```
num <- 4

if (num < 0) {
  num <- num * -1
}

num
## 4
```

The condition of an if statement must evaluate to a *single* TRUE or FALSE. If the condition creates a vector of TRUEs and FALSEs (which is easier to make than you may think), your if statement will print a warning message and use only the first element of the vector. Remember that you can condense vectors of logical values to a single TRUE or FALSE with the functions any and all.

You don't have to limit your if statements to a single line of code; you can include as many lines as you like between the braces. For example, the following code uses many lines to ensure that num is positive. The additional lines print some informative statements if num begins as a negative number. R will skip the entire code block—print statements and all—if num begins as a positive number:

```
num <- -1

if (num < 0) {
  print("num is negative.")
  print("Don't worry, I'll fix it.")
  num <- num * -1
```

```
    print("Now num is positive.")
}
## "num is negative."
## "Don't worry, I'll fix it."
## "Now num is positive."

num
## 1
```

Try the following quizzes to develop your understanding of if statements.

Quiz A

What will this return?

```
x <- 1
if (3 == 3) {
  x <- 2
}
x
```

Answer: the code will return the number 2. x begins as 1, and then R encounters the if statement. Since the condition evaluates to TRUE, R will run x <- 2, changing the value of x.

Quiz B

What will this return?

```
x <- 1
if (TRUE) {
  x <- 2
}
x
```

Answer: this code will also return the number 2. It works the same as the code in Quiz A, except the condition in this statement is already TRUE. R doesn't even need to evaluate it. As a result, the code inside the if statement will be run, and x will be set to 2.

Quiz C

What will this return?

```
x <- 1
if (x == 1) {
  x <- 2
  if (x == 1) {
```

```
      x <- 3
    }
  }
x
```

Answer: once again, the code will return the number 2. x starts out as 1, and the condition of the first if statement will evaluate to TRUE, which causes R to run the code in the body of the if statement. First, R sets x equal to 2, then R evaluates the second if statement, which is in the body of the first. This time x == 1 will evaluate to FALSE because x now equals 2. As a result, R ignores x <- 3 and exits both if statements.

else Statements

if statements tell R what to do when your condition is *true*, but you can also tell R what to do when the condition is *false*. else is a counterpart to if that extends an if statement to include a second case. In English, you would say, "If this is true, do plan A; else do plan B." In R, you would say:

```
if (this) {
   Plan A
} else {
   Plan B
}
```

When this evaluates to TRUE, R will run the code in the first set of braces, but not the code in the second. When this evaluates to FALSE, R will run the code in the second set of braces, but not the first. You can use this arrangement to cover all of the possible cases. For example, you could write some code that rounds a decimal to the nearest integer.

Start with a decimal:

```
a <- 3.14
```

Then isolate the decimal component with trunc:

```
dec <- a - trunc(a)   ❶ ❷
dec
## 0.14
```

❶ trunc takes a number and returns only the portion of the number that appears to the left of the decimal place (i.e., the integer part of the number).

❷ a - trunc(a) is a convenient way to return the decimal part of a.

Then use an if else tree to round the number (either up or down):

```
if (dec >= 0.5) {
   a <- trunc(a) + 1
```

```
} else {
  a <- trunc(a)
}

a
## 3
```

If your situation has more than two mutually exclusive cases, you can string multiple if and else statements together by adding a new if statement immediately after else. For example:

```
a <- 1
b <- 1

if (a > b) {
  print("A wins!")
} else if (a < b) {
  print("B wins!")
} else {
  print("Tie.")
}
## "Tie."
```

R will work through the if conditions until one evaluates to TRUE, then R will ignore any remaining if and else clauses in the tree. If no conditions evaluate to TRUE, R will run the final else statement.

If two if statements describe mutually exclusive events, it is better to join the if statements with an else if than to list them separately. This lets R ignore the second if statement whenever the first returns a TRUE, which saves work.

You can use if and else to link the subtasks in your slot-machine function. Open a fresh R script, and copy this code into it. The code will be the skeleton of our final score function. Compare it to the flow chart for score in Figure 7-2:

```
if ( # Case 1: all the same ❶ ) {
  prize <- # look up the prize ❷
} else if ( # Case 2: all bars ❸ ) {
  prize <- # assign $5 ❹
} else {
  # count cherries ❺
  prize <- # calculate a prize ❻
}

# count diamonds ❼
# double the prize if necessary ❽
```

Our skeleton is rather incomplete; there are many sections that are just code comments instead of real code. However, we've reduced the program to eight simple subtasks:

❶ Test whether the symbols are three of a kind.

❸ Look up the prize for three of a kind based on the common symbol.
❷ Test whether the symbols are all bars.
❹ Assign a prize of $5.
❺ Count the number of cherries.
❼ Count the number of diamonds.
❻ Calculate a prize based on the number of cherries.
❽ Adjust the prize for diamonds.

If you like, you can reorganize your flow chart around these tasks, as in Figure 7-4. The chart will describe the same strategy, but in a more precise way. I'll use a diamond shape to symbolize an `if else` decision.

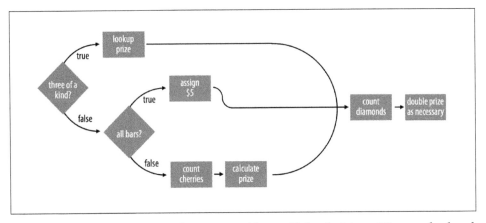

Figure 7-4. score can navigate three cases with two if else decisions. We can also break some of our tasks into two steps.

Now we can work through the subtasks one at a time, adding R code to the `if` tree as we go. Each subtask will be easy to solve if you set up a concrete example to work with and try to describe a solution in English before coding in R.

The first subtask asks you to test whether the symbols are three of a kind. How should you begin writing the code for this subtask?

You know that the final `score` function will look something like this:

```
score <- function(symbols) {

  # calculate a prize

  prize
}
```

Its argument, `symbols`, will be the output of `get_symbols`, a vector that contains three character strings. You could start writing `score` as I have written it, by defining an object named `score` and then slowly filling in the body of the function. However, this would be a bad idea. The eventual function will have eight separate parts, and it will not work correctly until *all* of those parts are written (and themselves work correctly). This means you would have to write the entire `score` function before you could test any of the subtasks. If `score` doesn't work—which is very likely—you will not know which subtask needs fixed.

You can save yourself time and headaches if you focus on one subtask at a time. For each subtask, create a concrete example that you can test your code on. For example, you know that `score` will need to work on a vector named `symbols` that contains three character strings. If you make a real vector named `symbols`, you can run the code for many of your subtasks on the vector as you go:

```
symbols <- c("7", "7", "7")
```

If a piece of code does not work on `symbols`, you will know that you need to fix it before you move on. You can change the value of `symbols` from subtask to subtask to ensure that your code works in every situation:

```
symbols <- c("B", "BB", "BBB")
symbols <- c("C", "DD", "0")
```

Only combine your subtasks into a `score` function once each subtask works on a concrete example. If you follow this plan, you will spend more time using your functions and less time trying to figure out why they do not work.

After you set up a concrete example, try to describe how you will do the subtask in English. The more precisely you can describe your solution, the easier it will be to write your R code.

Our first subtask asks us to "test whether the symbols are three of a kind." This phrase does not suggest any useful R code to me. However, I could describe a more precise test for three of a kind: three symbols will be the same if the first symbol is equal to the second and the second symbol is equal to the third. Or, even more precisely:

A vector named `symbols` will contain three of the same symbol if the first element of `symbols` is equal to the second element of `symbols` and the second element of `symbols` is equal to the third element of `symbols`.

Exercise

Turn the preceding statement into a logical test written in R. Use your knowledge of logical tests, Boolean operators, and subsetting from Chapter 4. The test should work with the vector symbols and return a TRUE *if and only if* each element in symbols is the same. Be sure to test your code on symbols.

Here are a couple of ways to test that symbols contains three of the same symbol. The first method parallels the English suggestion above, but there are other ways to do the same test. There is no right or wrong answer, so long as your solution works, which is easy to check because you've created a vector named symbols:

```
symbols
##  "7" "7" "7"

symbols[1] == symbols[2] & symbols[2] == symbols[3]
## TRUE

symbols[1] == symbols[2] & symbols[1] == symbols[3]
## TRUE

all(symbols == symbols[1])
## TRUE
```

As your vocabulary of R functions broadens, you'll think of more ways to do basic tasks. One method that I like for checking three of a kind is:

```
length(unique(symbols)) == 1
```

The unique function returns every unique term that appears in a vector. If your symbols vector contains three of a kind (i.e., one unique term that appears three times), then unique(symbols) will return a vector of length 1.

Now that you have a working test, you can add it to your slot-machine script:

```
same <- symbols[1] == symbols[2] && symbols[2] == symbols[3]  ❶

if (same) {
  prize <- # look up the prize
} else if ( # Case 2: all bars ) {
  prize <- # assign $5
} else {
  # count cherries
  prize <- # calculate a prize
}

# count diamonds
# double the prize if necessary
```

❶ && and || behave like & and | but can sometimes be more efficient. The double operators will not evaluate the second test in a pair of tests if the first test makes the result clear. For example, if symbols[1] does not equal symbols[2] in the next expression, && will not evaluate symbols[2] == symbols[3]; it can immediately return a FALSE for the whole expression (because FALSE & TRUE and FALSE & FALSE both evaluate to FALSE). This efficiency can speed up your programs; however, double operators are not appropriate everywhere. && and || are not vectorized, which means they can only handle a single logical test on each side of the operator.

The second prize case occurs when all the symbols are a type of bar, for example, B, BB, and BBB. Let's begin by creating a concrete example to work with:

```
symbols <- c("B", "BBB", "BB")
```

> **Exercise**
>
> Use R's logical and Boolean operators to write a test that will determine whether a vector named symbols contains only symbols that are a type of bar. Check whether your test works with our example symbols vector. Remember to describe how the test should work in English, and then convert the solution to R.

As with many things in R, there are multiple ways to test whether symbols contains all bars. For example, you could write a very long test that uses multiple Boolean operators, like this:

```
symbols[1] == "B" | symbols[1] == "BB" | symbols[1] == "BBB" &
  symbols[2] == "B" | symbols[2] == "BB" | symbols[2] == "BBB" &
  symbols[3] == "B" | symbols[3] == "BB" | symbols[3] == "BBB"
## TRUE
```

However, this is not a very efficient solution, because R has to run nine logical tests (and you have to type them). You can often replace multiple | operators with a single %in%. Also, you can check that a test is true for each element in a vector with all. These two changes shorten the preceding code to:

```
all(symbols %in% c("B", "BB", "BBB"))
## TRUE
```

Let's add this code to our script:

```
same <- symbols[1] == symbols[2] && symbols[2] == symbols[3]
bars <- symbols %in% c("B", "BB", "BBB")

if (same) {
  prize <- # look up the prize
```

```
} else if (all(bars)) {
  prize <- # assign $5
} else {
  # count cherries
  prize <- # calculate a prize
}

# count diamonds
# double the prize if necessary
```

You may have noticed that I split this test up into two steps, bars and all(bars). That's just a matter of personal preference. Wherever possible, I like to write my code so it can be read with function and object names conveying what they do.

You also may have noticed that our test for Case 2 will capture some symbols that should be in Case 1 because they contain three of a kind:

```
symbols <- c("B", "B", "B")
all(symbols %in% c("B", "BB", "BBB"))
## TRUE
```

That won't be a problem, however, because we've connected our cases with else if in the if tree. As soon as R comes to a case that evaluates to TRUE, it will skip over the rest of the tree. Think of it this way: each else tells R to only run the code that follows it *if none of the previous conditions have been met*. So when we have three of the same type of bar, R will evaluate the code for Case 1 and then skip the code for Case 2 (and Case 3).

Our next subtask is to assign a prize for symbols. When the symbols vector contains three of the same symbol, the prize will depend on which symbol there are three of. If there are three DDs, the prize will be $100; if there are three 7s, the prize will be $80; and so on.

This suggests another if tree. You could assign a prize with some code like this:

```
if (same) {
  symbol <- symbols[1]
  if (symbol == "DD") {
    prize <- 800
  } else if (symbol == "7") {
    prize <- 80
  } else if (symbol == "BBB") {
    prize <- 40
  } else if (symbol == "BB") {
    prize <- 5
  } else if (symbol == "B") {
    prize <- 10
  } else if (symbol == "C") {
    prize <- 10
  } else if (symbol == "0") {
    prize <- 0
```

```
    }
  }
```

While this code will work, it is a bit long to write and read, and it may require R to perform multiple logical tests before delivering the correct prize. We can do better with a different method.

Lookup Tables

Very often in R, the simplest way to do something will involve subsetting. How could you use subsetting here? Since you know the exact relationship between the symbols and their prizes, you can create a vector that captures this information. This vector can store symbols as names and prize values as elements:

```
payouts <- c("DD" = 100, "7" = 80, "BBB" = 40, "BB" = 25,
  "B" = 10, "C" = 10, "0" = 0)
payouts
##  DD   7 BBB  BB   B   C   0
## 100  80  40  25  10  10   0
```

Now you can extract the correct prize for any symbol by subsetting the vector with the symbol's name:

```
payouts["DD"]
##  DD
## 100

payouts["B"]
##  B
## 10
```

If you want to leave behind the symbol's name when subsetting, you can run the unname function on the output:

```
unname(payouts["DD"]) ❶
## 100
```

❶ unname returns a copy of an object with the names attribute removed.

payouts is a type of *lookup table*, an R object that you can use to look up values. Subsetting payouts provides a simple way to find the prize for a symbol. It doesn't take many lines of code, and it does the same amount of work whether your symbol is DD or 0. You can create lookup tables in R by creating named objects that can be subsetted in clever ways.

Sadly, our method is not quite automatic; we need to tell R which symbol to look up in payouts. Or do we? What would happen if you subsetted payouts by symbols[1]? Give it a try:

```
symbols <- c("7", "7", "7")
symbols[1]
## "7"

payouts[symbols[1]]
##  7
## 80

symbols <- c("C", "C", "C")
payouts[symbols[1]]
## C
## 10
```

You don't need to know the exact symbol to look up because you can tell R to look up whichever symbol happens to be in `symbols`. You can find this symbol with `symbols[1]`, `symbols[2]`, or `symbols[3]`, because each contains the same symbol in this case. You now have a simple automated way to calculate the prize when `symbols` contains three of a kind. Let's add it to our code and then look at Case 2:

```
same <- symbols[1] == symbols[2] && symbols[2] == symbols[3]
bars <- symbols %in% c("B", "BB", "BBB")

if (same) {
  payouts <- c("DD" = 100, "7" = 80, "BBB" = 40, "BB" = 25,
    "B" = 10, "C" = 10, "0" = 0)
  prize <- unname(payouts[symbols[1]])
} else if (all(bars)) {
  prize <- # assign $5
} else {
  # count cherries
  prize <- # calculate a prize
}

# count diamonds
# double the prize if necessary
```

Case 2 occurs whenever the symbols are all bars. In that case, the prize will be $5, which is easy to assign:

```
same <- symbols[1] == symbols[2] && symbols[2] == symbols[3]
bars <- symbols %in% c("B", "BB", "BBB")

if (same) {
  payouts <- c("DD" = 100, "7" = 80, "BBB" = 40, "BB" = 25,
    "B" = 10, "C" = 10, "0" = 0)
  prize <- unname(payouts[symbols[1]])
} else if (all(bars)) {
  prize <- 5
} else {
  # count cherries
  prize <- # calculate a prize
}
```

```
# count diamonds
# double the prize if necessary
```

Now we can work on the last case. Here, you'll need to know how many cherries are in `symbols` before you can calculate a prize.

> **Exercise**
>
> How can you tell which elements of a vector named `symbols` are a C? Devise a test and try it out.

> **Challenge**
>
> How might you count the number of Cs in a vector named `symbols`? Remember R's coercion rules.

As always, let's work with a real example:

```
symbols <- c("C", "DD", "C")
```

One way to test for cherries would be to check which, if any, of the symbols are a C:

```
symbols == "C"
## TRUE FALSE  TRUE
```

It'd be even more useful to count how many of the symbols are cherries. You can do this with `sum`, which expects numeric input, not logical. Knowing this, R will coerce the TRUEs and FALSEs to 1s and 0s before doing the summation. As a result, `sum` will return the number of TRUEs, which is also the number of cherries:

```
sum(symbols == "C")
## 2
```

You can use the same method to count the number of diamonds in `symbols`:

```
sum(symbols == "DD")
## 1
```

Let's add both of these subtasks to the program skeleton:

```
same <- symbols[1] == symbols[2] && symbols[2] == symbols[3]
bars <- symbols %in% c("B", "BB", "BBB")

if (same) {
  payouts <- c("DD" = 100, "7" = 80, "BBB" = 40, "BB" = 25,
    "B" = 10, "C" = 10, "0" = 0)
  prize <- unname(payouts[symbols[1]])
```

```
} else if (all(bars)) {
  prize <- 5
} else {
  cherries <- sum(symbols == "C")
  prize <- # calculate a prize
}

diamonds <- sum(symbols == "DD")
# double the prize if necessary
```

Since Case 3 appears further down the `if` tree than Cases 1 and 2, the code in Case 3 will only be applied to players that do not have three of a kind or all bars. According to the slot machine's payout scheme, these players will win $5 if they have two cherries and $2 if they have one cherry. If the player has no cherries, she gets a prize of $0. We don't need to worry about three cherries because that outcome is already covered in Case 1.

As in Case 1, you could write an `if` tree that handles each combination of cherries, but just like in Case 1, this would be an inefficient solution:

```
if (cherries == 2) {
  prize <- 5
} else if (cherries == 1) {
  prize <- 2
} else {}
  prize <- 0
}
```

Again, I think the best solution will involve subsetting. If you are feeling ambitious, you can try to work this solution out on your own, but you will learn just as quickly by mentally working through the following proposed solution.

We know that our prize should be $0 if we have no cherries, $2 if we have one cherry, and $5 if we have two cherries. You can create a vector that contains this information. This will be a very simple lookup table:

```
c(0, 2, 5)
```

Now, like in Case 1, you can subset the vector to retrieve the correct prize. In this case, the prize's aren't identified by a symbol name, but by the number of cherries present. Do we have that information? Yes, it is stored in `cherries`. We can use basic integer subsetting to get the correct prize from the prior lookup table, for example, `c(0, 2, 5)[1]`.

`cherries` isn't exactly suited for integer subsetting because it could contain a zero, but that's easy to fix. We can subset with `cherries + 1`. Now when `cherries` equals zero, we have:

```
cherries + 1
## 1
```

```
c(0, 2, 5)[cherries + 1]
## 0
```

When `cherries` equals one, we have:

```
cherries + 1
## 2

c(0, 2, 5)[cherries + 1]
## 2
```

And when `cherries` equals two, we have:

```
cherries + 1
## 3

c(0, 2, 5)[cherries + 1]
## 5
```

Examine these solutions until you are satisfied that they return the correct prize for each number of cherries. Then add the code to your script, as follows:

```
same <- symbols[1] == symbols[2] && symbols[2] == symbols[3]
bars <- symbols %in% c("B", "BB", "BBB")

if (same) {
  payouts <- c("DD" = 100, "7" = 80, "BBB" = 40, "BB" = 25,
    "B" = 10, "C" = 10, "0" = 0)
  prize <- unname(payouts[symbols[1]])
} else if (all(bars)) {
  prize <- 5
} else {
  cherries <- sum(symbols == "C")
  prize <- c(0, 2, 5)[cherries + 1]
}

diamonds <- sum(symbols == "DD")
# double the prize if necessary
```

Lookup Tables Versus if Trees

This is the second time we've created a lookup table to avoid writing an `if` tree. Why is this technique helpful and why does it keep appearing? Many `if` trees in R are essential. They provide a useful way to tell R to use different algorithms in different cases. However, `if` trees are not appropriate everywhere.

`if` trees have a couple of drawbacks. First, they require R to run multiple tests as it works its way down the `if` tree, which can create unnecessary work. Second, as you'll see in Chapter 10, it can be difficult to use `if` trees in vectorized code, a style of code that takes advantage of R's programming strengths to create fast programs. Lookup tables do not suffer from either of these drawbacks.

> You won't be able to replace every if tree with a lookup table, nor should you. However, you can usually use lookup tables to avoid assigning variables with if trees. As a general rule, use an if tree if each branch of the tree runs different *code*. Use a lookup table if each branch of the tree only assigns a different *value*.
>
> To convert an if tree to a lookup table, identify the values to be assigned and store them in a vector. Next, identify the selection criteria used in the conditions of the if tree. If the conditions use character strings, give your vector names and use name-based subsetting. If the conditions use integers, use integer-based subsetting.

The final subtask is to double the prize once for every diamond present. This means that the final prize will be some multiple of the current prize. For example, if no diamonds are present, the prize will be:

```
prize * 1     # 1 = 2 ^ 0
```

If one diamond is present, it will be:

```
prize * 2     # 2 = 2 ^ 1
```

If two diamonds are present, it will be:

```
prize * 4     # 4 = 2 ^ 2
```

And if three diamonds are present, it will be:

```
prize * 8     # 8 = 2 ^ 3
```

Can you think of an easy way to handle this? How about something similar to these examples?

> ### Exercise
>
> Write a method for adjusting prize based on diamonds. Describe a solution in English first, and then write your code.

Here is a concise solution inspired by the previous pattern. The adjusted prize will equal:

```
prize * 2 ^ diamonds
```

which gives us our final score script:

```
same <- symbols[1] == symbols[2] && symbols[2] == symbols[3]
bars <- symbols %in% c("B", "BB", "BBB")

if (same) {
  payouts <- c("DD" = 100, "7" = 80, "BBB" = 40, "BB" = 25,
    "B" = 10, "C" = 10, "0" = 0)
  prize <- unname(payouts[symbols[1]])
```

```
  } else if (all(bars)) {
    prize <- 5
  } else {
    cherries <- sum(symbols == "C")
    prize <- c(0, 2, 5)[cherries + 1]
  }

  diamonds <- sum(symbols == "DD")
  prize * 2 ^ diamonds
```

Code Comments

You now have a working score script that you can save to a function. Before you save your script, though, consider adding comments to your code with a #. Comments can make your code easier to understand by explaining *why* the code does what it does. You can also use comments to break long programs into scannable chunks. For example, I would include three comments in the score code:

```
# identify case
same <- symbols[1] == symbols[2] && symbols[2] == symbols[3]
bars <- symbols %in% c("B", "BB", "BBB")

# get prize
if (same) {
  payouts <- c("DD" = 100, "7" = 80, "BBB" = 40, "BB" = 25,
    "B" = 10, "C" = 10, "0" = 0)
  prize <- unname(payouts[symbols[1]])
} else if (all(bars)) {
  prize <- 5
} else {
  cherries <- sum(symbols == "C")
  prize <- c(0, 2, 5)[cherries + 1]
}

# adjust for diamonds
diamonds <- sum(symbols == "DD")
prize * 2 ^ diamonds
```

Now that each part of your code works, you can wrap it into a function with the methods you learned in "Writing Your Own Functions" on page 16. Either use RStudio's Extract Function option in the menu bar under Code, or use the function function. Ensure that the last line of the function returns a result (it does), and identify any arguments used by your function. Often the concrete examples that you used to test your code, like symbols, will become the arguments of your function. Run the following code to start using the score function:

```
score <- function (symbols) {
  # identify case
  same <- symbols[1] == symbols[2] && symbols[2] == symbols[3]
  bars <- symbols %in% c("B", "BB", "BBB")
```

```
  # get prize
  if (same) {
    payouts <- c("DD" = 100, "7" = 80, "BBB" = 40, "BB" = 25,
      "B" = 10, "C" = 10, "0" = 0)
    prize <- unname(payouts[symbols[1]])
  } else if (all(bars)) {
    prize <- 5
  } else {
    cherries <- sum(symbols == "C")
    prize <- c(0, 2, 5)[cherries + 1]
  }

  # adjust for diamonds
  diamonds <- sum(symbols == "DD")
  prize * 2 ^ diamonds
}
```

Once you have defined the score function, the play function will work as well:

```
play <- function() {
  symbols <- get_symbols()
  print(symbols)
  score(symbols)
}
```

Now it is easy to play the slot machine:

```
play()
## "0"   "BB"  "B"
## 0

play()
## "DD"  "0"   "B"
## 0

play()
## "BB"  "BB"  "BB"
## 25
```

Summary

An R program is a set of instructions for your computer to follow that has been organized into a sequence of steps and cases. This may make programs seem simple, but don't be fooled: you can create complicated results with the right combination of simple steps (and cases).

As a programmer, you are more likely to be fooled in the opposite way. A program may seem impossible to write when you know that it must do something impressive. Do not panic in these situations. Divide the job before you into simple tasks, and then divide the tasks again. You can visualize the relationship between tasks with a flow chart if it helps. Then work on the subtasks one at a time. Describe solutions in English, then convert them to R code. Test each solution against concrete examples as you go. Once each of your subtasks works, combine your code into a function that you can share and reuse.

R provides tools that can help you do this. You can manage cases with `if` and `else` statements. You can create a lookup table with objects and subsetting. You can add code comments with `#`. And you can save your programs as a function with `function`.

Things often go wrong when people write programs. It will be up to you to find the source of any errors that occur and to fix them. It should be easy to find the source of your errors if you use a stepwise approach to writing functions, writing—and then testing—one bit at a time. However, if the source of an error eludes you, or you find yourself working with large chunks of untested code, consider using R's built in debugging tools, described in Appendix E.

The next two chapters will teach you more tools that you can use in your programs. As you master these tools, you will find it easier to write R programs that let you do whatever you wish to your data. In Chapter 8, you will learn how to use R's S3 system, an invisible hand that shapes many parts of R. You will use the system to build a custom class for your slot machine output, and you will tell R how to display objects that have your class.

CHAPTER 8
S3

You may have noticed that your slot machine results do not look the way I promised they would. I suggested that the slot machine would display its results like this:

```
play()
## 0 0 DD
## $0
```

But the current machine displays its results in a less pretty format:

```
play()
## "0"   "0"  "DD"
## 0
```

Moreover, the slot machine uses a hack to display symbols (we call `print` from within `play`). As a result, the symbols do not follow your prize output if you save it:

```
one_play <- play()
## "B" "0" "B"

one_play
## 0
```

You can fix both of these problems with R's S3 system.

The S3 System

S3 refers to a class system built into R. The system governs how R handles objects of different classes. Certain R functions will look up an object's S3 class, and then behave differently in response.

The `print` function is like this. When you print a numeric vector, `print` will display a number:

```
num <- 1000000000
print(num)
## 1000000000
```

But if you give that number the S3 class POSIXct followed by POSIXt, print will display a time:

```
class(num) <- c("POSIXct", "POSIXt")
print(num)
## "2001-09-08 19:46:40 CST"
```

If you use objects with classes—and you do—you will run into R's S3 system. S3 behavior can seem odd at first, but is easy to predict once you are familiar with it.

R's S3 system is built around three components: attributes (especially the class attribute), generic functions, and methods.

Attributes

In "Attributes" on page 43, you learned that many R objects come with attributes, pieces of extra information that are given a name and appended to the object. Attributes do not affect the values of the object, but stick to the object as a type of metadata that R can use to handle the object. For example, a data frame stores its row and column names as attributes. Data frames also store their class, "data.frame", as an attribute.

You can see an object's attributes with attribute. If you run attribute on the DECK data frame that you created in Part II, you will see:

```
attributes(DECK)
## $names
## [1] "face"  "suit"  "value"
##
## $class
## [1] "data.frame"
##
## $row.names
##  [1]  1  2  3  4  5  6  7  8  9 10 11 12 13 14 15 16 17 18 19
## [20] 20 21 22 23 24 25 26 27 28 29 30 31 32 33 34 35 36
## [37] 37 38 39 40 41 42 43 44 45 46 47 48 49 50 51 52
```

R comes with many helper functions that let you set and access the most common attributes used in R. You've already met the names, dim, and class functions, which each work with an eponymously named attribute. However, R also has row.names, levels, and many other attribute-based helper functions. You can use any of these functions to retrieve an attribute's value:

```
row.names(DECK)
##  [1] "1"  "2"  "3"  "4"  "5"  "6"  "7"  "8"  "9"  "10" "11" "12" "13"
## [14] "14" "15" "16" "17" "18" "19" "20" "21" "22" "23" "24" "25" "26"
```

```
## [27] "27" "28" "29" "30" "31" "32" "33" "34" "35" "36" "37" "38" "39"
## [40] "40" "41" "42" "43" "44" "45" "46" "47" "48" "49" "50" "51" "52"
```

or to change an attribute's value:

```
row.names(DECK) <- 101:152
```

or to give an object a new attribute altogether:

```
levels(DECK) <- c("level 1", "level 2", "level 3")

attributes(DECK)
## $names
## [1] "face"  "suit"  "value"
##
## $class
## [1] "data.frame"
##
## $row.names
##  [1] 101 102 103 104 105 106 107 108 109 110 111 112 113 114 115 116 117
## [18] 118 119 120 121 122 123 124 125 126 127 128 129 130 131 132 133 134
## [35] 135 136 137 138 139 140 141 142 143 144 145 146 147 148 149 150 151
## [52] 152
##
## $levels
## [1] "level 1" "level 2" "level 3"
```

R is very laissez faire when it comes to attributes. It will let you add any attributes that you like to an object (and then it will usually ignore them). The only time R will complain is when a function needs to find an attribute and it is not there.

You can add any general attribute to an object with `attr`; you can also use `attr` to look up the value of any attribute of an object. Let's see how this works with `one_play`, the result of playing our slot machine one time:

```
one_play <- play()
one_play
## 0

attributes(one_play)
## NULL
```

`attr` takes two arguments: an R object and the name of an attribute (as a character string). To give the R object an attribute of the specified name, save a value to the output of `attr`. Let's give `one_play` an attribute named `symbols` that contains a vector of character strings:

```
attr(one_play, "symbols") <- c("B", "0", "B")

attributes(one_play)
## $symbols
## [1] "B" "0" "B"
```

To look up the value of any attribute, give `attr` an R object and the name of the attribute you would like to look up:

```
attr(one_play, "symbols")
## "B" "0" "B"
```

If you give an attribute to an atomic vector, like `one_play`, R will usually display the attribute beneath the vector's values. However, if the attribute changes the vector's class, R may display all of the information in the vector in a new way (as we saw with POSIXct objects):

```
one_play
## [1] 0
## attr(,"symbols")
## [1] "B" "0" "B"
```

R will generally ignore an object's attributes unless you give them a name that an R function looks for, like `names` or `class`. For example, R will ignore the `symbols` attribute of `one_play` as you manipulate `one_play`:

```
one_play + 1
## 1
## attr(,"symbols")
##    "B" "0" "B"
```

Exercise

Modify `play` to return a prize that contains the symbols associated with it as an attribute named `symbols`. Remove the redundant call to `print(symbols)`:

```
play <- function() {
  symbols <- get_symbols()
  print(symbols)
  score(symbols)
}
```

You can create a new version of `play` by capturing the output of `score(symbols)` and assigning an attribute to it. `play` can then return the enhanced version of the output:

```
play <- function() {
  symbols <- get_symbols()
  prize <- score(symbols)
  attr(prize, "symbols") <- symbols
  prize
}
```

Now `play` returns both the prize and the symbols associated with the prize. The results may not look pretty, but the symbols stick with the prize when we copy it to a new object. We can work on tidying up the display in a minute:

```
play()
## [1] 0
## attr(,"symbols")
## [1] "B"  "BB" "0"

two_play <- play()

two_play
## [1] 0
## attr(,"symbols")
## [1] "0" "B" "0"
```

You can also generate a prize and set its attributes in one step with the `structure` function. `structure` creates an object with a set of attributes. The first argument of `structure` should be an R object or set of values, and the remaining arguments should be named attributes for `structure` to add to the object. You can give these arguments any argument names you like. `structure` will add the attributes to the object under the names that you provide as argument names:

```
play <- function() {
  symbols <- get_symbols()
  structure(score(symbols), symbols = symbols)
}

three_play <- play()
three_play
##   0
## attr(,"symbols")
## "0"  "BB" "B"
```

Now that your `play` output contains a `symbols` attribute, what can you do with it? You can write your own functions that lookup and use the attribute. For example, the following function will look up one_play's `symbols` attribute and use it to display one_play in a pretty manner. We will use this function to display our slot results, so let's take a moment to study what it does:

```
slot_display <- function(prize){

  # extract symbols
  symbols <- attr(prize, "symbols")

  # collapse symbols into single string
  symbols <- paste(symbols, collapse = " ")

  # combine symbol with prize as a regular expression
  # \n is regular expression for new line (i.e. return or enter)
  string <- paste(symbols, prize, sep = "\n$")

  # display regular expression in console without quotes
  cat(string)
}
```

```
slot_display(one_play)
## B 0 B
## $0
```

The function expects an object like `one_play` that has both a numerical value and a `symbols` attribute. The first line of the function will look up the value of the `symbols` attribute and save it as an object named `symbols`. Let's make an example `symbols` object so we can see what the rest of the function does. We can use `one_play`'s `symbols` attribute to do the job. `symbols` will be a vector of three-character strings:

```
symbols <- attr(one_play, "symbols")

symbols
## "B" "0" "B"
```

Next, `slot_display` uses `paste` to collapse the three strings in `symbols` into a single-character string. `paste` collapses a vector of character strings into a single string when you give it the `collapse` argument. `paste` will use the value of `collapse` to separate the formerly distinct strings. Hence, `symbols` becomes B 0 B the three strings separated by a space:

```
symbols <- paste(symbols, collapse = " ")

symbols
## "B 0 B"
```

Our function then uses `paste` in a new way to combine `symbols` with the value of `prize`. `paste` combines separate objects into a character string when you give it a `sep` argument. For example, here `paste` will combine the string in `symbols`, B 0 B, with the number in `prize`, 0. `paste` will use the value of `sep` argument to separate the inputs in the new string. Here, that value is `\n$`, so our result will look like `"B 0 B\n$0"`:

```
prize <- one_play
string <- paste(symbols, prize, sep = "\n$")

string
## "B 0 B\n$0"
```

The last line of `slot_display` calls `cat` on the new string. `cat` is like `print`; it displays its input at the command line. However, `cat` does not surround its output with quotation marks. `cat` also replaces every `\n` with a new line or line break. The result is what we see. Notice that it looks just how I suggested that our `play` output should look in Chapter 7:

```
cat(string)
## B 0 B
## $0
```

You can use `slot_display` to manually clean up the output of `play`:

```
slot_display(play())
## C B 0
## $2

slot_display(play())
## 7 0 BB
## $0
```

This method of cleaning the output requires you to manually intervene in your R session (to call `slot_display`). There is a function that you can use to automatically clean up the output of `play` *each* time it is displayed. This function is `print`, and it is a *generic function*.

Generic Functions

R uses `print` more often than you may think; R calls `print` each time it displays a result in your console window. This call happens in the background, so you do not notice it; but the call explains how output makes it to the console window (recall that `print` always prints its argument in the console window). This `print` call also explains why the output of `print` always matches what you see when you display an object at the command line:

```
print(pi)
## 3.141593

pi
## 3.141593


print(head(deck))
##    face   suit value
##    king spades    13
##   queen spades    12
##    jack spades    11
##     ten spades    10
##    nine spades     9
##   eight spades     8

head(deck)
##    face   suit value
##    king spades    13
##   queen spades    12
##    jack spades    11
##     ten spades    10
##    nine spades     9
##   eight spades     8


print(play())
## 5
## attr(,"symbols")
```

```
##  "B"  "BB" "B" 
play()
##  5
## attr(,"symbols")
##  "B"  "BB" "B" 
```

You can change how R displays your slot output by rewriting `print` to look like `slot_display`. Then R would print the output in our tidy format. However, this method would have negative side effects. You do not want R to call `slot_display` when it prints a data frame, a numerical vector, or any other object.

Fortunately, `print` is not a normal function; it is a *generic* function. This means that `print` is written in a way that lets it do different things in different cases. You've already seen this behavior in action (although you may not have realized it). `print` did one thing when we looked at the unclassed version of `num`:

```
num <- 1000000000
print(num)
## 1000000000
```

and a different thing when we gave `num` a class:

```
class(num) <- c("POSIXct", "POSIXt")
print(num)
## "2001-09-08 19:46:40 CST"
```

Take a look at the code inside `print` to see how it does this. You may imagine that print looks up the class attribute of its input and then uses an `if` tree to pick which output to display. If this occurred to you, great job! `print` does something very similar, but much more simple.

Methods

When you call `print`, `print` calls a special function, `UseMethod`:

```
print
## function (x, ...)
## UseMethod("print")
## <bytecode: 0x7ffee4c62f80>
## <environment: namespace:base>
```

`UseMethod` examines the class of the input that you provide for the first argument of `print`, and then passes all of your arguments to a new function designed to handle that class of input. For example, when you give `print` a POSIXct object, `UseMethod` will pass all of `print`'s arguments to `print.POSIXct`. R will then run `print.POSIXct` and return the results:

```
print.POSIXct
## function (x, ...)
```

```
##  {
##      max.print <- getOption("max.print", 9999L)
##      if (max.print < length(x)) {
##          print(format(x[seq_len(max.print)], usetz = TRUE), ...)
##          cat(" [ reached getOption(\"max.print\") -- omitted",
##              length(x) - max.print, "entries ]\n")
##      }
##      else print(format(x, usetz = TRUE), ...)
##      invisible(x)
##  }
## <bytecode: 0x7fa948f3d008>
## <environment: namespace:base>
```

If you give print a factor object, UseMethod will pass all of print's arguments to print.factor. R will then run print.factor and return the results:

```
print.factor
## function (x, quote = FALSE, max.levels = NULL, width = getOption("width"),
##     ...)
## {
##     ord <- is.ordered(x)
##     if (length(x) == 0L)
##         cat(if (ord)
##             "ordered"
## ...
##         drop <- n > maxl
##         cat(if (drop)
##             paste(format(n), ""), T0, paste(if (drop)
##             c(lev[1L:max(1, maxl - 1)], "..."), if (maxl > 1) lev[n])
##         else lev, collapse = colsep), "\n", sep = "")
##     }
##     invisible(x)
## }
## <bytecode: 0x7fa94a64d470>
## <environment: namespace:base>
```

print.POSIXct and print.factor are called *methods* of print. By themselves, print.POSIXct and print.factor work like regular R functions. However, each was written specifically so UseMethod could call it to handle a specific class of print input.

Notice that print.POSIXct and print.factor do two different things (also notice that I abridged the middle of print.factor — it is a long function). This is how print manages to do different things in different cases. print calls UseMethod, which calls a specialized method based on the class of print's first argument.

You can see which methods exist for a generic function by calling methods on the function. For example, print has almost 200 methods (which gives you an idea of how many classes exist in R):

```
methods(print)
##   [1] print.acf*
```

```
##     [2] print.anova
##     [3] print.aov*
##     ...
## [176] print.xgettext*
## [177] print.xngettext*
## [178] print.xtabs*
##
##   Nonvisible functions are asterisked
```

This system of generic functions, methods, and class-based dispatch is known as S3 because it originated in the third version of S, the programming language that would evolve into S-PLUS and R. Many common R functions are S3 generics that work with a set of class methods. For example, `summary` and `head` also call `UseMethod`. More basic functions, like `c`, `+`, `-`, `<` and others also behave like generic functions, although they call `.primitive` instead of `UseMethod`.

The S3 system allows R functions to behave in different ways for different classes. You can use S3 to format your slot output. First, give your output its own class. Then write a print method for that class. To do this efficiently, you will need to know a little about how `UseMethod` selects a method function to use.

Method Dispatch

`UseMethod` uses a very simple system to match methods to functions.

Every S3 method has a two-part name. The first part of the name will refer to the function that the method works with. The second part will refer to the class. These two parts will be separated by a period. So for example, the print method that works with functions will be called `print.function`. The summary method that works with matrices will be called `summary.matrix`. And so on.

When `UseMethod` needs to call a method, it searches for an R function with the correct S3-style name. The function does not have to be special in any way; it just needs to have the correct name.

You can participate in this system by writing your own function and giving it a valid S3-style name. For example, let's give `one_play` a class of its own. It doesn't matter what you call the class; R will store any character string in the class attribute:

```
class(one_play) <- "slots"
```

Now let's write an S3 print method for the `slots` class. The method doesn't need to do anything special—it doesn't even need to print `one_play`. But it *does* need to be named `print.slots`; otherwise `UseMethod` will not find it. The method should also take the same arguments as `print`; otherwise, R will give an error when it tries to pass the arguments to `print.slots`:

```
args(print)
## function (x, ...)
## NULL

print.slots <- function(x, ...) {
  cat("I'm using the print.slots method")
}
```

Does our method work? Yes, and not only that; R uses the print method to display the contents of one_play. This method isn't very useful, so I'm going to remove it. You'll have a chance to write a better one in a minute:

```
print(one_play)
## I'm using the print.slots method

one_play
## I'm using the print.slots method

rm(print.slots)
```

Some R objects have multiple classes. For example, the output of Sys.time has two classes. Which class will UseMethod use to find a print method?

```
now <- Sys.time()
attributes(now)
## $class
## [1] "POSIXct" "POSIXt"
```

UseMethod will first look for a method that matches the first class listed in the object's class vector. If UseMethod cannot find one, it will then look for the method that matches the second class (and so on if there are more classes in an object's class vector).

If you give print an object whose class or classes do not have a print method, UseMethod will call print.default, a special method written to handle general cases.

Let's use this system to write a better print method for the slot machine output.

Exercise

Write a new print method for the slots class. The method should call slot_display to return well-formatted slot-machine output.

What name must you use for this method?

It is surprisingly easy to write a good print.slots method because we've already done all of the hard work when we wrote slot_display. For example, the following method will work. Just make sure the method is named print.slots so UseMethod can find it, and make sure that it takes the same arguments as `print` so `UseMethod` can pass those arguments to `print.slots` without any trouble:

```
print.slots <- function(x, ...) {
  slot_display(x)
}
```

Now R will automatically use `slot_display` to display objects of class `slots` (and only objects of class "slots"):

```
one_play
## B 0 B
## $0
```

Let's ensure that every piece of slot machine output has the `slots` class.

Exercise

Modify the `play` function so it assigns `slots` to the `class` attribute of its output:

```
play <- function() {
  symbols <- get_symbols()
  structure(score(symbols), symbols = symbols)
}
```

You can set the `class` attribute of the output at the same time that you set the `symbols` attribute. Just add `class = "slots"` to the `structure` call:

```
play <- function() {
  symbols <- get_symbols()
  structure(score(symbols), symbols = symbols, class = "slots")
}
```

Now each of our slot machine plays will have the class `slots`:

```
class(play())
## "slots"
```

As a result, R will display them in the correct slot-machine format:

```
play()
## BB BB BBB
## $5

play()
## BB 0 0
## $0
```

Classes

You can use the S3 system to make a robust new class of objects in R. Then R will treat objects of your class in a consistent, sensible manner. To make a class:

1. Choose a name for your class.
2. Assign each instance of your class a class attribute.
3. Write class methods for any generic function likely to use objects of your class.

Many R packages are based on classes that have been built in a similar manner. While this work is simple, it may not be easy. For example, consider how many methods exist for predefined classes.

You can call methods on a class with the class argument, which takes a character string. methods will return every method written for the class. Notice that methods will not be able to show you methods that come in an unloaded R package:

```
methods(class = "factor")
##  [1] [.factor            [[.factor
##  [3] [[<-.factor         [<-.factor
##  [5] all.equal.factor    as.character.factor
##  [7] as.data.frame.factor as.Date.factor
##  [9] as.list.factor      as.logical.factor
## [11] as.POSIXlt.factor   as.vector.factor
## [13] droplevels.factor   format.factor
## [15] is.na<-.factor      length<-.factor
## [17] levels<-.factor     Math.factor
## [19] Ops.factor          plot.factor*
## [21] print.factor        relevel.factor*
## [23] relist.factor*      rep.factor
## [25] summary.factor      Summary.factor
## [27] xtfrm.factor
##
##    Nonvisible functions are asterisked
```

This output indicates how much work is required to create a robust, well-behaved class. You will usually need to write a class method for every basic R operation.

Consider two challenges that you will face right away. First, R drops attributes (like class) when it combines objects into a vector:

```
play1 <- play()
play1
## B BBB BBB
## $5

play2 <- play()
play2
## 0 B 0
```

```
## $0

c(play1, play2)
## [1] 5 0
```

Here, R stops using `print.slots` to display the vector because the vector `c(play1, play2)` no longer has a "slots" `class` attribute.

Next, R will drop the attributes of an object (like `class`) when you subset the object:

```
play1[1]
## [1] 5
```

You can avoid this behavior by writing a `c.slots` method and a `[.slots` method, but then difficulties will quickly accrue. How would you combine the `symbols` attributes of multiple plays into a vector of symbols attributes? How would you change `print.slots` to handle vectors of outputs? These challenges are open for you to explore. However, you will usually not have to attempt this type of large-scale programming as a data scientist.

In our case, it is very handy to let `slots` objects revert to single prize values when we combine groups of them together into a vector.

S3 and Debugging

S3 can be annoying if you are trying to understand R functions. It is difficult to tell what a function does if its code body contains a call to `UseMethod`. Now that you know that `UseMethod` calls a class-specific method, you can search for and examine the method directly. It will be a function whose name follows the `<function.class>` syntax, or possibly `<function.default>`. You can also use the `methods` function to see what methods are associated with a function or a class.

S4 and R5

R also contains two other systems that create class specific behavior. These are known as S4 and R5 (or reference classes). Each of these systems is much harder to use than S3, and perhaps as a consequence, more rare. However, they offer safeguards that S3 does not. If you'd like to learn more about these systems, including how to write and use your own generic functions, I recommend the forthcoming book *Advanced R Programming* by Hadley Wickham.

Summary

Values are not the only place to store information in R, and functions are not the only way to create unique behavior. You can also do both of these things with R's S3 system.

The S3 system provides a simple way to create object-specific behavior in R. In other words, it is R's version of object-oriented programming (OOP). The system is implemented by generic functions. These functions examine the class attribute of their input and call a class-specific method to generate output. Many S3 methods will look for and use additional information that is stored in an object's attributes. Many common R functions are S3 generics.

R's S3 system is more helpful for the tasks of computer science than the tasks of data science, but understanding S3 can help you troubleshoot your work in R as a data scientist.

You now know quite a bit about how to write R code that performs custom tasks, but how could you repeat these tasks? As a data scientist, you will often repeat tasks, sometimes thousands or even millions of times. Why? Because repetition lets you simulate results and estimate probabilities. Chapter 9 will show you how to automate repetition with R's for and while functions. You'll use for to simulate various slot machine plays and to calculate the payout rate of your slot machine.

CHAPTER 9
Loops

Loops are R's method for repeating a task, which makes them a useful tool for programming simulations. This chapter will teach you how to use R's loop tools.

Let's use the score function to solve a real-world problem.

Your slot machine is modeled after real machines that were accused of fraud. The machines appeared to pay out 40 cents on the dollar, but the manufacturer claimed that they paid out 92 cents on the dollar. You can calculate the exact payout rate of your machine with the score program. The payout rate will be the expected value of the slot machine's prize.

Expected Values

The expected value of a random event is a type of weighted average; it is the sum of each possible outcome of the event, weighted by the probability that each outcome occurs:

$$E(x) = \sum_{i=1}^{n} (x_i \cdot P(x_i))$$

You can think of the expected value as the average prize that you would observe if you played the slot machine an infinite number of times. Let's use the formula to calculate some simple expected values. Then we will apply the formula to your slot machine.

Do you remember the die you created in Part I?

```
die <- c(1, 2, 3, 4, 5, 6)
```

Each time you roll the die, it returns a value selected at random (one through six). You can find the expected value of rolling the die with the formula:

$$E(\text{die}) = \sum_{i=1}^{n} (\text{die}_i \cdot P(\text{die}_i))$$

The die_is are the possible outcomes of rolling the die: 1, 2, 3, 4, 5, and 6; and the $P(\text{die}_i)$'s are the probabilities associated with each of the outcomes. If your die is fair, each outcome will occur with the same probability: 1/6. So our equation simplifies to:

$$\begin{aligned} E(\text{die}) &= \sum_{i=1}^{n} (\text{die}_i \cdot P(\text{die}_i)) \\ &= 1 \cdot \frac{1}{6} + 2 \cdot \frac{1}{6} + 3 \cdot \frac{1}{6} + 4 \cdot \frac{1}{6} + 5 \cdot \frac{1}{6} + 6 \cdot \frac{1}{6} \\ &= 3.5 \end{aligned}$$

Hence, the expected value of rolling a fair die is 3.5. You may notice that this is also the average value of the die. The expected value will equal the average if every outcome has the same chance of occurring.

But what if each outcome has a different chance of occurring? For example, we weighted our dice in Chapter 2 so that each die rolled 1, 2, 3, 4, and 5 with probability 1/8 and 6 with probability 3/8. You can use the same formula to calculate the expected value in these conditions:

$$\begin{aligned} E(\text{die}) &= 1 \cdot \frac{1}{8} + 2 \cdot \frac{1}{8} + 3 \cdot \frac{1}{8} + 4 \cdot \frac{1}{8} + 5 \cdot \frac{1}{8} + 6 \cdot \frac{3}{8} \\ &= 4.125 \end{aligned}$$

Hence, the expected value of a loaded die does not equal the average value of its outcomes. If you rolled a loaded die an infinite number of times, the average outcome would be 4.125, which is higher than what you would expect from a fair die.

Notice that we did the same three things to calculate both of these expected values. We have:

1. Listed out all of the possible outcomes
2. Determined the *value* of each outcome (here just the value of the die)
3. Calculated the probability that each outcome occurred

The expected value was then just the sum of the values in step 2 multiplied by the probabilities in step 3.

You can use these steps to calculate more sophisticated expected values. For example, you could calculate the expected value of rolling a pair of weighted dice. Let's do this step by step.

First, list out all of the possible outcomes. A total of 36 different outcomes can appear when you roll two dice. For example, you might roll (1, 1), which notates one on the first die and one on the second die. Or, you may roll (1, 2), one on the first die and two on the second. And so on. Listing out these combinations can be tedious, but R has a function that can help.

expand.grid

The `expand.grid` function in R provides a quick way to write out every combination of the elements in *n* vectors. For example, you can list every combination of two dice. To do so, run `expand.grid` on two copies of `die`:

```
rolls <- expand.grid(die, die)
```

`expand.grid` will return a data frame that contains every way to pair an element from the first `die` vector with an element from the second `die` vector. This will capture all 36 possible combinations of values:

```
rolls
##    Var1 Var2
## 1    1    1
## 2    2    1
## 3    3    1
## ...
## 34   4    6
## 35   5    6
## 36   6    6
```

You can use `expand.grid` with more than two vectors if you like. For example, you could list every combination of rolling three dice with `expand.grid(die, die, die)` and every combination of rolling four dice with `expand.grid(die, die, die, die)`, and so on. `expand.grid` will always return a data frame that contains each possible combination of *n* elements from the *n* vectors. Each combination will contain exactly one element from each vector.

You can determine the value of each roll once you've made your list of outcomes, . This will be the sum of the two dice, which you can calculate using R's element-wise execution:

```
rolls$value <- rolls$Var1 + rolls$Var2
head(rolls, 3)
## Var1 Var2 value
##   1    1    2
##   2    1    3
##   3    1    4
```

R will match up the elements in each vector before adding them together. As a result, each element of `value` will refer to the elements of `Var1` and `Var2` that appear in the same row.

Next, you must determine the probability that each combination appears. You can calculate this with a basic rule of probability:

The probability that n independent, random events all occur is equal to the product of the probabilities that each random event occurs.

Or more succinctly:

$$P(A \ \& \ B \ \& \ C \ \& \ ...) = P(A) \cdot P(B) \cdot P(C) \cdot ...$$

So the probability that we roll a (1, 1) will be equal to the probability that we roll a one on the first die, 1/8, times the probability that we roll a one on the second die, 1/8:

$$\begin{aligned} P(1 \ \& \ 1) &= P(1) \cdot P(1) \\ &= \frac{1}{8} \cdot \frac{1}{8} \\ &= \frac{1}{64} \end{aligned}$$

And the probability that we roll a (1, 2) will be:

$$\begin{aligned} P(1 \ \& \ 2) &= P(1) \cdot P(2) \\ &= \frac{1}{8} \cdot \frac{1}{8} \\ &= \frac{1}{64} \end{aligned}$$

And so on.

Let me suggest a three-step process for calculating these probabilities in R. First, we can look up the probabilities of rolling the values in `Var1`. We'll do this with the lookup table that follows:

```
prob <- c("1" = 1/8, "2" = 1/8, "3" = 1/8, "4" = 1/8, "5" = 1/8, "6" = 3/8)

prob
##     1     2     3     4     5     6
## 0.125 0.125 0.125 0.125 0.125 0.375
```

If you subset this table by `rolls$Var1`, you will get a vector of probabilities perfectly keyed to the values of `Var1`:

```
rolls$Var1
## 1 2 3 4 5 6 1 2 3 4 5 6 1 2 3 4 5 6 1 2 3 4 5 6 1 2 3 4 5 6 1 2 3 4 5 6

prob[rolls$Var1]
```

```
## 1     2     3     4     5     6     1     2     3     4     5     6
## 0.125 0.125 0.125 0.125 0.125 0.375 0.125 0.125 0.125 0.125 0.125 0.375
## 1     2     3     4     5     6     1     2     3     4     5     6
## 0.125 0.125 0.125 0.125 0.125 0.375 0.125 0.125 0.125 0.125 0.125 0.375
## 1     2     3     4     5     6     1     2     3     4     5     6
## 0.125 0.125 0.125 0.125 0.125 0.375 0.125 0.125 0.125 0.125 0.125 0.375

rolls$prob1 <- prob[rolls$Var1]
head(rolls, 3)
## Var1 Var2 value prob1
##   1    1     2   0.125
##   2    1     3   0.125
##   3    1     4   0.125
```

Second, we can look up the probabilities of rolling the values in Var2:

```
rolls$prob2 <- prob[rolls$Var2]

head(rolls, 3)
## Var1 Var2 value prob1 prob2
##   1    1     2   0.125 0.125
##   2    1     3   0.125 0.125
##   3    1     4   0.125 0.125
```

Third, we can calculate the probability of rolling each combination by multiplying prob1 by prob2:

```
rolls$prob <- rolls$prob1 * rolls$prob2

head(rolls, 3)
## Var1 Var2 value prob1 prob2   prob
##   1    1     2   0.125 0.125 0.015625
##   2    1     3   0.125 0.125 0.015625
##   3    1     4   0.125 0.125 0.015625
```

It is easy to calculate the expected value now that we have each outcome, the value of each outcome, and the probability of each outcome. The expected value will be the summation of the dice values multiplied by the dice probabilities:

```
sum(rolls$value * rolls$prob)
## 8.25
```

So the expected value of rolling two loaded dice is 8.25. If you rolled a pair of loaded dice an infinite number of times, the average sum would be 8.25. (If you are curious, the expected value of rolling a pair of fair dice is 7, which explains why 7 plays such a large role in dice games like craps.)

Now that you've warmed up, let's use our method to calculate the expected value of the slot machine prize. We will follow the same steps we just took:

1. We will list out every possible outcome of playing the machine. This will be a list of every combination of three slot symbols.

2. We will calculate the probability of getting each combination when you play the machine.
3. We will determine the prize that we would win for each combination.

When we are finished, we will have a data set that looks like this:

```
## Var1 Var2 Var3 prob1 prob2 prob3      prob prize
##   DD   DD   DD  0.03  0.03  0.03 0.000027   800
##    7   DD   DD  0.03  0.03  0.03 0.000027     0
##  BBB   DD   DD  0.06  0.03  0.03 0.000054     0
## ... and so on.
```

The expected value will then be the sum of the prizes multiplied by their probability of occuring:

$$E(\text{prize}) = \sum_{i=1}^{n} (\text{prize}_i \cdot P(\text{prize}_i))$$

Ready to begin?

> **Exercise**
>
> Use `expand.grid` to make a data frame that contains every possible combination of *three* symbols from the `wheel` vector:
>
> ```
> wheel <- c("DD", "7", "BBB", "BB", "B", "C", "0")
> ```
>
> Be sure to add the argument `stringsAsFactors = FALSE` to your `expand.grid` call; otherwise, `expand.grid` will save the combinations as factors, an unfortunate choice that will disrupt the `score` function.

To create a data frame of each combination of *three* symbols, you need to run `expand.grid` and give it *three* copies of `wheel`. The result will be a data frame with 343 rows, one for each unique combination of three slot symbols:

```
combos <- expand.grid(wheel, wheel, wheel, stringsAsFactors = FALSE)
combos
##     Var1 Var2 Var3
## 1     DD   DD   DD
## 2      7   DD   DD
## 3    BBB   DD   DD
## 4     BB   DD   DD
## 5      B   DD   DD
## 6      C   DD   DD
## ...
## 341    B    0    0
```

```
## 342    C    0    0
## 343    0    0    0
```

Now, let's calculate the probability of getting each combination. You can use the probabilities contained in the prob argument of get_symbols to do this. These probabilities determine how frequently each symbol is chosen when your slot machine generates symbols. They were calculated after observing 345 plays of the Manitoba video lottery terminals. Zeroes have the largest chance of being selected (0.52) and cherries the least (0.01):

```
get_symbols <- function() {
  wheel <- c("DD", "7", "BBB", "BB", "B", "C", "0")
  sample(wheel, size = 3, replace = TRUE,
    prob = c(0.03, 0.03, 0.06, 0.1, 0.25, 0.01, 0.52))
}
```

> **Exercise**
>
> Isolate the previous probabilities in a lookup table. What names will you use in your table?

Your names should match the input that you want to look up. In this case, the input will be the character strings that appear in Var1, Var2, and Var3. So your lookup table should look like this:

```
prob <- c("DD" = 0.03, "7" = 0.03, "BBB" = 0.06,
  "BB" = 0.1, "B" = 0.25, "C" = 0.01, "0" = 0.52)
```

Now let's look up our probabilities.

> **Exercise**
>
> Look up the probabilities of getting the values in Var1. Then add them to combos as a column named prob1. Then do the same for Var2 (prob2) and Var3 (prob3).

Remember that you use R's selection notation to look up values in a lookup table. The values that result will be keyed to the index that you use:

```
combos$prob1 <- prob[combos$Var1]
combos$prob2 <- prob[combos$Var2]
combos$prob3 <- prob[combos$Var3]

head(combos, 3)
## Var1 Var2 Var3 prob1 prob2 prob3
##   DD   DD   DD  0.03  0.03  0.03
```

```
##   7   DD  DD  0.03  0.03  0.03
## BBB   DD  DD  0.06  0.03  0.03
```

Now how should we calculate the total probability of each combination? Our three slot symbols are all chosen independently, which means that the same rule that governed our dice probabilities governs our symbol probabilities:

$P(A \text{ \& } B \text{ \& } C \text{ \& } ...) = P(A) \cdot P(B) \cdot P(C) \cdot ...$

Exercise

Calculate the overall probabilities for each combination. Save them as a column named prob in combos, then check your work.

You can check that the math worked by summing the probabilities. The probabilities should add up to one, because one of the combinations *must* appear when you play the slot machine. In other words, a combination will appear, with probability of one.

You can calculate the probabilities of every possible combination in one fell swoop with some element-wise execution:

```
combos$prob <- combos$prob1 * combos$prob2 * combos$prob3

head(combos, 3)
## Var1 Var2 Var3 prob1 prob2 prob3     prob
##   DD   DD   DD  0.03  0.03  0.03 0.000027
##    7   DD   DD  0.03  0.03  0.03 0.000027
##  BBB   DD   DD  0.06  0.03  0.03 0.000054
```

The sum of the probabilities is one, which suggests that our math is correct:

```
sum(combos$prob)
## 1
```

You only need to do one more thing before you can calculate the expected value: you must determine the prize for each combination in combos. You can calculate the prize with score. For example, we can calculate the prize for the first row of combos like this:

```
symbols <- c(combos[1, 1], combos[1, 2], combos[1, 3])
## "DD" "DD" "DD"

score(symbols)
## 800
```

However there are 343 rows, which makes for tedious work if you plan to calculate the scores manually. It will be quicker to automate this task and have R do it for you, which you can do with a for loop.

for Loops

A for loop repeats a chunk of code many times, once for each element in a set of input. for loops provide a way to tell R, "Do this for every value of that." In R syntax, this looks like:

```
for (value in that) {
  this
}
```

The that object should be a set of objects (often a vector of numbers or character strings). The for loop will run the code in that appears between the braces once for each member of that. For example, the for loop below runs print("one run") once for each element in a vector of character strings:

```
for (value in c("My", "first", "for", "loop")) {
  print("one run")
}
## "one run"
## "one run"
## "one run"
## "one run"
```

The value symbol in a for loop acts like an argument in a function. The for loop will create an object named value and assign it a new value on each run of the loop. The code in your loop can access this value by calling the value object.

What values will the for loop assign to value? It will use the elements in the set that you run the loop on. for starts with the first element and then assigns a different element to value on each run of the for loop, until all of the elements have been assigned to value. For example, the for loop below will run print(value) four times and will print out one element of c("My", "second", "for", "loop") each time:

```
for (value in c("My", "second", "for", "loop")) {
  print(value)
}
## "My"
## "second"
## "for"
## "loop"
```

On the first run, the for loop substituted "My" for value in print(value). On the second run it substituted "second", and so on until for had run print(value) once with every element in the set.

If you look at value after the loop runs, you will see that it still contains the value of the last element in the set:

```
value
## "loop"
```

I've been using the symbol value in my for loops, but there is nothing special about it. You can use any symbol you like in your loop to do the same thing as long as the symbol appears before in in the parentheses that follow for. For example, you could rewrite the previous loop with any of the following:

```
for (word in c("My", "second", "for", "loop")) {
  print(word)
}
for (string in c("My", "second", "for", "loop")) {
  print(string)
}
for (i in c("My", "second", "for", "loop")) {
  print(i)
}
```

Choose your symbols carefully

R will run your loop in whichever environment you call it from. This is bad news if your loop uses object names that already exist in the environment. Your loop will overwrite the existing objects with the objects that it creates. This applies to the value symbol as well.

For loops run on sets

In many programming languages, for loops are designed to work with integers, not sets. You give the loop a starting value and an ending value, as well as an increment to advance the value by between loops. The for loop then runs until the loop value exceeds the ending value.

You can recreate this effect in R by having a for loop execute on a set of integers, but don't lose track of the fact that R's for loops execute on members of a set, not sequences of integers.

for loops are very useful in programming because they help you connect a piece of code with each element in a set. For example, we could use a for loop to run score once for each row in combos. However, R's for loops have a shortcoming that you'll want to know about before you start using them: for loops do not return output.

for loops are like Las Vegas: what happens in a for loop stays in a for loop. If you want to use the products of a for loop, you must write the for loop so that it saves its own output as it goes.

Our previous examples appeared to return output, but this was misleading. The examples worked because we called print, which always prints its arguments in the console (even if it is called from a function, a for loop, or anything else). Our for loops won't return anything if you remove the print call:

```
for (value in c("My", "third", "for", "loop")) {
  value
}
##
```

To save output from a for loop, you must write the loop so that it saves its own output as it runs. You can do this by creating an empty vector or list before you run the for loop. Then use the for loop to fill up the vector or list. When the for loop is finished, you'll be able to access the vector or list, which will now have all of your results.

Let's see this in action. The following code creates an empty vector of length 4:

```
chars <- vector(length = 4)
```

The next loop will fill it with strings:

```
words <- c("My", "fourth", "for", "loop")

for (i in 1:4) {
  chars[i] <- words[i]
}

chars
## "My"     "fourth" "for"    "loop"
```

This approach will usually require you to change the sets that you execute your for loop on. Instead of executing on a set of objects, execute on a set of integers that you can use to index both your object and your storage vector. This approach is very common in R. You'll find in practice that you use for loops not so much to run code, but to fill up vectors and lists with the results of code.

Let's use a for loop to calculate the prize for each row in combos. To begin, create a new column in combos to store the results of the for loop:

```
combos$prize <- NA

head(combos, 3)
##   Var1 Var2 Var3 prob1 prob2 prob3      prob prize
##     DD   DD   DD  0.03  0.03  0.03  0.000027    NA
##      7   DD   DD  0.03  0.03  0.03  0.000027    NA
##    BBB   DD   DD  0.06  0.03  0.03  0.000054    NA
```

The code creates a new column named prize and fills it with NAs. R uses its recycling rules to populate every value of the column with NA.

> **Exercise**
>
> Construct a `for` loop that will run `score` on all 343 rows of `combos`. The loop should run `score` on the first three entries of the *i*th row of `combos` and should store the results in the *i*th entry of `combos$prize`.

You can score the rows in `combos` with:

```
for (i in 1:nrow(combos)) {
  symbols <- c(combos[i, 1], combos[i, 2], combos[i, 3])
  combos$prize[i] <- score(symbols)
}
```

After you run the for loop, `combos$prize` will contain the correct prize for each row. This exercise also tests the `score` function; `score` appears to work correctly for every possible slot combination:

```
head(combos, 3)
##   Var1 Var2 Var3 prob1 prob2 prob3     prob prize
##   DD   DD   DD   0.03  0.03  0.03 0.000027   800
##    7   DD   DD   0.03  0.03  0.03 0.000027     0
##   BBB  DD   DD   0.06  0.03  0.03 0.000054     0
```

We're now ready to calculate the expected value of the prize. The expected value is the sum of `combos$prize` weighted by `combos$prob`. This is also the payout rate of the slot machine:

```
sum(combos$prize * combos$prob)
## 0.538014
```

Uh oh. The expected prize is about 0.54, which means our slot machine only pays 54 cents on the dollar over the long run. Does this mean that the manufacturer of the Manitoba slot machines *was* lying?

No, because we ignored an important feature of the slot machine when we wrote `score`: a diamond is wild. You can treat a `DD` as any other symbol if it increases your prize, with one exception. You cannot make a `DD` a `C` unless you already have another `C` in your symbols (it'd be too easy if every `DD` automatically earned you $2).

The best thing about `DD`s is that their effects are cumulative. For example, consider the combination B, DD, B. Not only does the `DD` count as a B, which would earn a prize of $10; the `DD` also doubles the prize to $20.

Adding this behavior to our code is a little tougher than what we have done so far, but it involves all of the same principles. You can decide that your slot machine doesn't use wilds and keep the code that we have. In that case, your slot machine will have a payout rate of about 54 percent. Or, you could rewrite your code to use wilds. If you do, you will find that your slot machine has a payout rate of 93 percent, one percent higher than the manufacturer's claim. You can calculate this rate with the same method that we used in this section.

> **Challenge**
>
> There are many ways to modify `score` that would count DDs as wild. If you would like to test your skill as an R programmer, try to write your own version of `score` that correctly handles diamonds.
>
> If you would like a more modest challenge, study the following `score` code. It accounts for wild diamonds in a way that I find elegant and succinct. See if you can understand each step in the code and how it achieves its result.

Here is a version of score that handles wild diamonds:

```
score <- function(symbols) {

  diamonds <- sum(symbols == "DD")
  cherries <- sum(symbols == "C")

  # identify case
  # since diamonds are wild, only nondiamonds
  # matter for three of a kind and all bars
  slots <- symbols[symbols != "DD"]
  same <- length(unique(slots)) == 1
  bars <- slots %in% c("B", "BB", "BBB")

  # assign prize
  if (diamonds == 3) {
    prize <- 100
  } else if (same) {
    payouts <- c("7" = 80, "BBB" = 40, "BB" = 25,
      "B" = 10, "C" = 10, "0" = 0)
    prize <- unname(payouts[slots[1]])
  } else if (all(bars)) {
    prize <- 5
  } else if (cherries > 0) {
    # diamonds count as cherries
    # so long as there is one real cherry
    prize <- c(0, 2, 5)[cherries + diamonds + 1]
  } else {
    prize <- 0
  }

  # double for each diamond
  prize * 2^diamonds
}
```

> **Exercise**
>
> Calculate the expected value of the slot machine when it uses the new `score` function. You can use the existing `combos` data frame, but you will need to build a `for` loop to recalculate `combos$prize`.

To update the expected value, just update `combos$prize`:

```
for (i in 1:nrow(combos)) {
  symbols <- c(combos[i, 1], combos[i, 2], combos[i, 3])
  combos$prize[i] <- score(symbols)
}
```

Then recompute the expected value:

```
sum(combos$prize * combos$prob)
## 0.934356
```

This result vindicates the manufacturer's claim. If anything, the slot machines seem more generous than the manufacturer stated.

while Loops

R has two companions to the `for` loop: the `while` loop and the `repeat` loop. A `while` loop reruns a chunk *while* a certain condition remains TRUE. To create a `while` loop, follow `while` by a condition and a chunk of code, like this:

```
while (condition) {
  code
}
```

`while` will rerun `condition`, which should be a logical test, at the start of each loop. If `condition` evaluates to TRUE, `while` will run the code between its braces. If `condition` evaluates to FALSE, `while` will finish the loop.

Why might `condition` change from TRUE to FALSE? Presumably because the code inside your loop has changed whether the condition is still TRUE. If the code has no relationship to the condition, a `while` loop will run until you stop it. So be careful. You can stop a `while` loop by hitting Escape or by clicking on the stop-sign icon at the top of the RStudio console pane. The icon will appear once the loop begins to run.

Like `for` loops, `while` loops do not return a result, so you must think about what you want the loop to return and save it to an object during the loop.

You can use `while` loops to do things that take a varying number of iterations, like calculating how long it takes to go broke playing slots (as follows). However, in practice, `while` loops are much less common than `for` loops in R:

```
plays_till_broke <- function(start_with) {
  cash <- start_with
  n <- 0
  while (cash > 0) {
    cash <- cash - 1 + play()
    n <- n + 1
  }
  n
}

plays_till_broke(100)
## 260
```

repeat Loops

repeat loops are even more basic than while loops. They will repeat a chunk of code until you tell them to stop (by hitting Escape) or until they encounter the command break, which will stop the loop.

You can use a repeat loop to recreate plays_till_broke, my function that simulates how long it takes to lose money while playing slots:

```
plays_till_broke <- function(start_with) {
  cash <- start_with
  n <- 0
  repeat {
    cash <- cash - 1 + play()
    n <- n + 1
    if (cash <= 0) {
      break
    }
  }
  n
}

plays_till_broke(100)
## 237
```

Summary

You can repeat tasks in R with for, while, and repeat loops. To use for, give it a chunk of code to run and a set of objects to loop through. for will run the code chunk once for each object. If you wish to save the output of your loop, you can assign it to an object that exists outside of the loop.

Repetition plays an important role in data science. It is the basis for simulation, as well as for estimates of variance and probability. Loops are not the only way to create repetition in R (consider replicate for example), but they are one of the most popular ways.

Unfortunately, loops in R can sometimes be slower than loops in other languages. As a result, R's loops get a bad rap. This reputation is not entirely deserved, but it does highlight an important issue. Speed is essential to data analysis. When your code runs fast, you can work with bigger data and do more to it before you run out of time or computational power. Chapter 10 will teach you how to write fast `for` loops and fast code in general with R. There, you will learn to write vectorized code, a style of lightning-fast code that takes advantage of all of R's strengths.

CHAPTER 10
Speed

As a data scientist, you need speed. You can work with bigger data and do more ambitious tasks when your code runs fast. This chapter will show you a specific way to write fast code in R. You will then use the method to simulate 10 million plays of your slot machine.

Vectorized Code

You can write a piece of code in many different ways, but the fastest R code will usually take advantage of three things: logical tests, subsetting, and element-wise execution. These are the things that R does best. Code that uses these things usually has a certain quality: it is *vectorized*; the code can take a vector of values as input and manipulate each value in the vector at the same time.

To see what vectorized code looks like, compare these two examples of an absolute value function. Each takes a vector of numbers and transforms it into a vector of absolute values (e.g., positive numbers). The first example is not vectorized; abs_loop uses a for loop to manipulate each element of the vector one at a time:

```
abs_loop <- function(vec){
  for (i in 1:length(vec)) {
    if (vec[i] < 0) {
      vec[i] <- -vec[i]
    }
  }
  vec
}
```

The second example, abs_set, is a vectorized version of abs_loop. It uses logical subsetting to manipulate every negative number in the vector at the same time:

```
abs_set <- function(vec){
  negs <- vec < 0
```

```
    vec[negs] <- vec[negs] * -1
    vec
}
```

`abs_set` is much faster than `abs_loop` because it relies on operations that R does quickly: logical tests, subsetting, and element-wise execution.

You can use the `system.time` function to see just how fast `abs_set` is. `system.time` takes an R expression, runs it, and then displays how much time elapsed while the expression ran.

To compare `abs_loop` and `abs_set`, first make a long vector of positive and negative numbers. `long` will contain 10 million values:

```
long <- rep(c(-1, 1), 5000000)  ❶
```

❶ `rep` repeats a value, or vector of values, many times. To use `rep`, give it a vector of values and then the number of times to repeat the vector. R will return the results as a new, longer vector.

You can then use `system.time` to measure how much time it takes each function to evaluate `long`:

```
system.time(abs_loop(long))  ❶
##    user  system elapsed
##  15.982   0.032  16.018

system.time(abs_set(long))
##    user  system elapsed
##   0.529   0.063   0.592
```

❶ Don't confuse `system.time` with `Sys.time`, which returns the current time.

The first two columns of the output of `system.time` report how many seconds your computer spent executing the call on the user side and system sides of your process, a dichotomy that will vary from OS to OS.

The last column displays how many seconds elapsed while R ran the expression. The results show that `abs_set` calculated the absolute value 30 times faster than `abs_loop` when applied to a vector of 10 million numbers. You can expect similar speed-ups whenever you write vectorized code.

Exercise

Many preexisting R functions are already vectorized and have been optimized to perform quickly. You can make your code faster by relying on these functions whenever possible. For example, R comes with a built-in absolute value function, `abs`.

Check to see how much faster abs computes the absolute value of long than abs_loop and abs_set do.

You can measure the speed of abs with system.time. It takes abs a lightning-fast 0.05 seconds to calculate the absolute value of 10 million numbers. This is 0.592 / 0.054 = 10.96 times faster than abs_set and nearly 300 times faster than abs_loop:

```
system.time(abs(long))
##    user  system elapsed
##   0.037   0.018   0.054
```

How to Write Vectorized Code

Vectorized code is easy to write in R because most R functions are already vectorized. Code based on these functions can easily be made vectorized and therefore fast. To create vectorized code:

1. Use vectorized functions to complete the sequential steps in your program.
2. Use logical subsetting to handle parallel cases. Try to manipulate every element in a case at once.

abs_loop and abs_set illustrate these rules. The functions both handle two cases and perform one sequential step, Figure 10-1. If a number is positive, the functions leave it alone. If a number is negative, the functions multiply it by negative one.

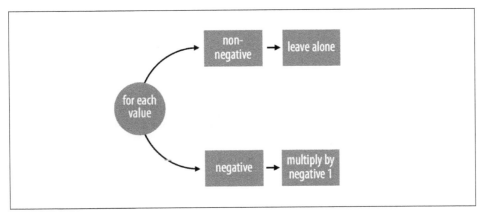

Figure 10-1. abs_loop uses a for loop to sift data into one of two cases: negative numbers and nonnegative numbers.

You can identify all of the elements of a vector that fall into a case with a logical test. R will execute the test in element-wise fashion and return a TRUE for every element that belongs in the case. For example, vec < 0 identifies every value of vec that belongs to the negative case. You can use the same logical test to extract the set of negative values with logical subsetting:

```
vec <- c(1, -2, 3, -4, 5, -6, 7, -8, 9, -10)
vec < 0
## FALSE TRUE FALSE TRUE FALSE TRUE FALSE TRUE FALSE TRUE

vec[vec < 0]
## -2  -4  -6  -8 -10
```

The plan in Figure 10-1 now requires a sequential step: you must multiply each of the negative values by negative one. All of R's arithmetic operators are vectorized, so you can use * to complete this step in vectorized fashion. * will multiply each number in vec[vec < 0] by negative one at the same time:

```
vec[vec < 0] * -1
## 2 4 6 8 10
```

Finally, you can use R's assignment operator, which is also vectorized, to save the new set over the old set in the original vec object. Since <- is vectorized, the elements of the new set will be paired up to the elements of the old set, in order, and then element-wise assignment will occur. As a result, each negative value will be replaced by its positive partner, as in Figure 10-2.

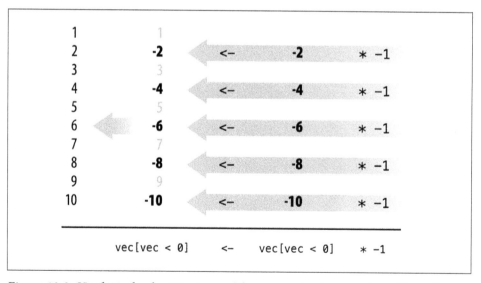

Figure 10-2. Use logical subsetting to modify groups of values in place. R's arithmetic and assignment operators are vectorized, which lets you manipulate and update multiple values at once.

Exercise

The following function converts a vector of slot symbols to a vector of new slot symbols. Can you vectorize it? How much faster does the vectorized version work?

```
change_symbols <- function(vec){
  for (i in 1:length(vec)){
    if (vec[i] == "DD") {
      vec[i] <- "joker"
    } else if (vec[i] == "C") {
      vec[i] <- "ace"
    } else if (vec[i] == "7") {
      vec[i] <- "king"
    }else if (vec[i] == "B") {
      vec[i] <- "queen"
    } else if (vec[i] == "BB") {
      vec[i] <- "jack"
    } else if (vec[i] == "BBB") {
      vec[i] <- "ten"
    } else {
      vec[i] <- "nine"
    }
  }
  vec
}

vec <- c("DD", "C", "7", "B", "BB", "BBB", "0")

change_symbols(vec)
##  "joker" "ace"   "king"  "queen" "jack"  "ten"   "nine"

many <- rep(vec, 1000000)

system.time(change_symbols(many))
##    user  system elapsed
##  30.057   0.031  30.079
```

`change_symbols` uses a `for` loop to sort values into seven different cases, as demonstrated in Figure 10-3.

To vectorize `change_symbols`, create a logical test that can identify each case:

```
vec[vec == "DD"]
## "DD"

vec[vec == "C"]
## "C"

vec[vec == "7"]
## "7"
```

```
vec[vec == "B"]
## "B"

vec[vec == "BB"]
## "BB"

vec[vec == "BBB"]
## "BBB"

vec[vec == "0"]
## "0"
```

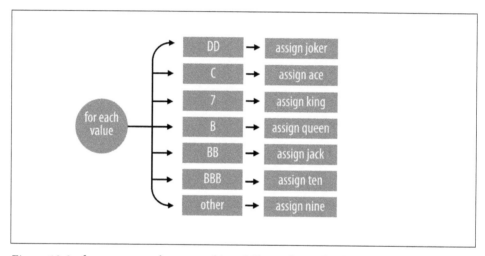

Figure 10-3. change_many does something different for each of seven cases.

Then write code that can change the symbols for each case:

```
vec[vec == "DD"] <- "joker"
vec[vec == "C"] <- "ace"
vec[vec == "7"] <- "king"
vec[vec == "B"] <- "queen"
vec[vec == "BB"] <- "jack"
vec[vec == "BBB"] <- "ten"
vec[vec == "0"] <- "nine"
```

When you combine this into a function, you have a vectorized version of change_sym
bols that runs about 14 times faster:

```
change_vec <- function (vec) {
  vec[vec == "DD"] <- "joker"
  vec[vec == "C"] <- "ace"
  vec[vec == "7"] <- "king"
  vec[vec == "B"] <- "queen"
  vec[vec == "BB"] <- "jack"
```

```
  vec[vec == "BBB"] <- "ten"
  vec[vec == "0"] <- "nine"

  vec
}

system.time(change_vec(many))
##   user system elapsed
##  1.994  0.059   2.051
```

Or, even better, use a lookup table. Lookup tables are a vectorized method because they rely on R's vectorized selection operations:

```
change_vec2 <- function(vec){
  tb <- c("DD" = "joker", "C" = "ace", "7" = "king", "B" = "queen",
    "BB" = "jack", "BBB" = "ten", "0" = "nine")
  unname(tb[vec])
}

system.time(change_vec(many))
##   user system elapsed
##  0.687  0.059   0.746
```

Here, a lookup table is 40 times faster than the original function.

`abs_loop` and `change_many` illustrate a characteristic of vectorized code: programmers often write slower, nonvectorized code by relying on unnecessary `for` loops, like the one in `change_many`. I think this is the result of a general misunderstanding about R. `for` loops do not behave the same way in R as they do in other languages, which means you should write code differently in R than you would in other languages.

When you write in languages like C and Fortran, you must compile your code before your computer can run it. This compilation step optimizes how the `for` loops in the code use your computer's memory, which makes the `for` loops very fast. As a result, many programmers use `for` loops frequently when they write in C and Fortran.

When you write in R, however, you do not compile your code. You skip this step, which makes programming in R a more user-friendly experience. Unfortunately, this also means you do not give your loops the speed boost they would receive in C or Fortran. As a result, your loops will run slower than the other operations we have studied: logical tests, subsetting, and element-wise execution. If you can write your code with the faster operations instead of a `for` loop, you should do so. No matter which language you write in, you should try to use the features of the language that run the fastest.

if and for

A good way to spot for loops that could be vectorized is to look for combinations of if and for. if can only be applied to one value at a time, which means it is often used in conjunction with a for loop. The for loop helps apply if to an entire vector of values. This combination can usually be replaced with logical subsetting, which will do the same thing but run much faster.

This doesn't mean that you should never use for loops in R. There are still many places in R where for loops make sense. for loops perform a basic task that you cannot always recreate with vectorized code. for loops are also easy to understand and run reasonably fast in R, so long as you take a few precautions.

How to Write Fast for Loops in R

You can dramatically increase the speed of your for loops by doing two things to optimize each loop. First, do as much as you can outside of the for loop. Every line of code that you place inside of the for loop will be run many, many times. If a line of code only needs to be run once, place it outside of the loop to avoid repetition.

Second, make sure that any storage objects that you use with the loop are large enough to contain *all* of the results of the loop. For example, both loops below will need to store one million values. The first loop stores its values in an object named output that begins with a length of *one million*:

```
system.time(
  output <- rep(NA, 1000000)
  for (i in 1:1000000) {
    output[i] <- i + 1
  }
})
)
##    user  system elapsed
##   1.709   0.015   1.724
```

The second loop stores its values in an object named output that begins with a length of *one*. R will expand the object to a length of one million as it runs the loop. The code in this loop is very similar to the code in the first loop, but the loop takes *37 minutes* longer to run than the first loop:

```
system.time(
  output <- NA
  for (i in 1:1000000) {
    output[i] <- i + 1
  }
)
```

180 | Chapter 10: Speed

```
##    user  system elapsed
## 1689.537 560.951 2249.927
```

The two loops do the same thing, so what accounts for the difference? In the second loop, R has to increase the length of output by one for each run of the loop. To do this, R needs to find a new place in your computer's memory that can contain the larger object. R must then copy the output vector over and erase the old version of output before moving on to the next run of the loop. By the end of the loop, R has rewritten output in your computer's memory one million times.

In the first case, the size of output never changes; R can define one output object in memory and use it for each run of the for loop.

The authors of R use low-level languages like C and Fortran to write basic R functions, many of which use for loops. These functions are compiled and optimized before they become a part of R, which makes them quite fast.

Whenever you see .Primitive, .Internal, or .Call written in a function's definition, you can be confident the function is calling code from another language. You'll get all of the speed advantages of that language by using the function.

Vectorized Code in Practice

To see how vectorized code can help you as a data scientist, consider our slot machine project. In Chapter 9, you calculated the exact payout rate for your slot machine, but you could have estimated this payout rate with a simulation. If you played the slot machine many, many times, the average prize over all of the plays would be a good estimate of the true payout rate.

This method of estimation is based on the law of large numbers and is similar to many statistical simulations. To run this simulation, you could use a for loop:

```
winnings <- vector(length = 1000000)
for (i in 1:1000000) {
  winnings[i] <- play()
}

mean(winnings)
## 0.9366984
```

The estimated payout rate after 10 million runs is 0.937, which is very close to the true payout rate of 0.934. Note that I'm using the modified score function that treats diamonds as wilds.

If you run this simulation, you will notice that it takes a while to run. In fact, the simulation takes 342.308 seconds to run, which is about 5.7 minutes. This is not particularly impressive, and you can do better by using vectorized code:

```
system.time(for (i in 1:1000000) {
  winnings[i] <- play()
})
##    user  system elapsed
## 342.041   0.355 342.308
```

The current score function is not vectorized. It takes a single slot combination and uses an if tree to assign a prize to it. This combination of an if tree with a for loop suggests that you could write a piece of vectorized code that takes *many* slot combinations and then uses logical subsetting to operate on them all at once.

For example, you could rewrite get_symbols to generate n slot combinations and return them as an $n \times 3$ matrix, like the one that follows. Each row of the matrix will contain one slot combination to be scored:

```
get_many_symbols <- function(n) {
  wheel <- c("DD", "7", "BBB", "BB", "B", "C", "0")
  vec <- sample(wheel, size = 3 * n, replace = TRUE,
    prob = c(0.03, 0.03, 0.06, 0.1, 0.25, 0.01, 0.52))
  matrix(vec, ncol = 3)
}

get_many_symbols(5)
##      [,1]  [,2] [,3]
## [1,] "B"   "0"  "B"
## [2,] "0"   "BB" "7"
## [3,] "0"   "0"  "BBB"
## [4,] "0"   "0"  "B"
## [5,] "BBB" "0"  "0"
```

You could also rewrite play to take a parameter, n, and return n prizes, in a data frame:

```
play_many <- function(n) {
  symb_mat <- get_many_symbols(n = n)
  data.frame(w1 = symb_mat[,1], w2 = symb_mat[,2],
             w3 = symb_mat[,3], prize = score_many(symb_mat))
}
```

This new function would make it easy to simulate a million, or even 10 million plays of the slot machine, which will be our goal. When we're finished, you will be able to estimate the payout rate with:

```
# plays <- play_many(10000000))
# mean(plays$prize)
```

Now you just need to write score_many, a vectorized (matrix-ized?) version of score that takes an $n \times 3$ matrix and returns n prizes. It will be difficult to write this function because score is already quite complicated. I would not expect you to feel confident doing this on your own until you have more practice and experience than we've been able to develop here.

Should you like to test your skills and write a version of score_many, I recommend that you use the function rowSums within your code. It calculates the sum of each row of numbers (or logicals) in a matrix.

If you would like to test yourself in a more modest way, I recommend that you study the following model score_many function until you understand how each part works and how the parts work together to create a vectorized function. To do this, it will be helpful to create a concrete example, like this:

```
symbols <- matrix(
  c("DD", "DD", "DD",
    "C", "DD", "0",
    "B", "B", "B",
    "B", "BB", "BBB",
    "C", "C", "0",
    "7", "DD", "DD"), nrow = 6, byrow = TRUE)

symbols
##      [,1] [,2] [,3]
## [1,] "DD" "DD" "DD"
## [2,] "C"  "DD" "0"
## [3,] "B"  "B"  "B"
## [4,] "B"  "BB" "BBB"
## [5,] "C"  "C"  "0"
## [6,] "7"  "DD" "DD"
```

Then you can run each line of score_many against the example and examine the results as you go.

Exercise

Study the model score_many function until you are satisfied that you understand how it works and could write a similar function yourself.

Advanced Challenge

Instead of examining the model answer, write your own vectorized version of score. Assume that the data is stored in an $n \times 3$ matrix where each row of the matrix contains one combination of slots to be scored.

You can use the version of score that treats diamonds as wild or the version of score that doesn't. However, the model answer will use the version treating diamonds as wild.

`score_many` is a vectorized version of `score`. You can use it to run the simulation at the start of this section in a little over 20 seconds. This is 17 times faster than using a for loop:

```
# symbols should be a matrix with a column for each slot machine window
score_many <- function(symbols) {

  # Step 1: Assign base prize based on cherries and diamonds ---------
  ## Count the number of cherries and diamonds in each combination
  cherries <- rowSums(symbols == "C")
  diamonds <- rowSums(symbols == "DD")

  ## Wild diamonds count as cherries
  prize <- c(0, 2, 5)[cherries + diamonds + 1]

  ## ...but not if there are zero real cherries
  ### (cherries is coerced to FALSE where cherries == 0)
  prize[!cherries] <- 0

  # Step 2: Change prize for combinations that contain three of a kind
  same <- symbols[, 1] == symbols[, 2] &
    symbols[, 2] == symbols[, 3]
  payoffs <- c("DD" = 100, "7" = 80, "BBB" = 40,
    "BB" = 25, "B" = 10, "C" = 10, "0" = 0)
  prize[same] <- payoffs[symbols[same, 1]]

  # Step 3: Change prize for combinations that contain all bars ------
  bars <- symbols == "B" | symbols == "BB" | symbols == "BBB"
  all_bars <- bars[, 1] & bars[, 2] & bars[, 3] & !same
  prize[all_bars] <- 5

  # Step 4: Handle wilds --------------------------------------------

  ## combos with two diamonds
  two_wilds <- diamonds == 2

  ### Identify the nonwild symbol
  one <- two_wilds & symbols[, 1] != symbols[, 2] &
    symbols[, 2] == symbols[, 3]
  two <- two_wilds & symbols[, 1] != symbols[, 2] &
    symbols[, 1] == symbols[, 3]
  three <- two_wilds & symbols[, 1] == symbols[, 2] &
    symbols[, 2] != symbols[, 3]

  ### Treat as three of a kind
  prize[one] <- payoffs[symbols[one, 1]]
  prize[two] <- payoffs[symbols[two, 2]]
  prize[three] <- payoffs[symbols[three, 3]]

  ## combos with one wild
  one_wild <- diamonds == 1
```

```
    ### Treat as all bars (if appropriate)
    wild_bars <- one_wild & (rowSums(bars) == 2)
    prize[wild_bars] <- 5

    ### Treat as three of a kind (if appropriate)
    one <- one_wild & symbols[, 1] == symbols[, 2]
    two <- one_wild & symbols[, 2] == symbols[, 3]
    three <- one_wild & symbols[, 3] == symbols[, 1]
    prize[one] <- payoffs[symbols[one, 1]]
    prize[two] <- payoffs[symbols[two, 2]]
    prize[three] <- payoffs[symbols[three, 3]]

    # Step 5: Double prize for every diamond in combo ------------------
    unname(prize * 2^diamonds)

}
system.time(play_many(10000000))
##    user  system elapsed
## 20.942   1.433  22.367
```

Loops Versus Vectorized Code

In many languages, for loops run very fast. As a result, programmers learn to use for loops whenever possible when they code. Often these programmers continue to rely on for loops when they begin to program in R, usually without taking the simple steps needed to optimize R's for loops. These programmers may become disillusioned with R when their code does not work as fast as they would like. If you think that this may be happening to you, examine how often you are using for loops and what you are using them to do. If you find yourself using for loops for every task, there is a good chance that you are "speaking R with a C accent." The cure is to learn to write and use vectorized code.

This doesn't mean that for loops have no place in R. for loops are a very useful feature; they can do many things that vectorized code cannot do. You also should not become a slave to vectorized code. Sometimes it would take more time to rewrite code in vectorized format than to let a for loop run. For example, would it be faster to let the slot simulation run for 5.7 minutes or to rewrite score?

Summary

Fast code is an important component of data science because you can do more with fast code than you can do with slow code. You can work with larger data sets before computational constraints intervene, and you can do more computation before time constraints intervene. The fastest code in R will rely on the things that R does best: logical tests, subsetting, and element-wise execution. I've called this type of code vectorized code because code written with these operations will take a vector of values as input and operate on each element of the vector at the same time. The majority of the code written in R is already vectorized.

If you use these operations, but your code does not appear vectorized, analyze the sequential steps and parallel cases in your program. Ensure that you've used vectorized functions to handle the steps and logical subsetting to handle the cases. Be aware, however, that some tasks cannot be vectorized.

Project 3 Wrap-up

You have now written your first program in R, and it is a program that you should be proud of. `play` is not a simple `hello world` exercise, but a real program that does a real task in a complicated way.

Writing new programs in R will always be challenging because programming depends so much on your own creativity, problem-solving ability, and experience writing similar types of programs. However, you can use the suggestions in this chapter to make even the most complicated program manageable: divide tasks into simple steps and cases, work with concrete examples, and describe possible solutions in English.

This project completes the education you began in Chapter 1. You can now use R to handle data, which has augmented your ability to analyze data. You can:

- Load and store data in your computer—not on paper or in your mind
- Accurately recall and change individual values without relying on your memory
- Instruct your computer to do tedious, or complex, tasks on your behalf

These skills solve an important logistical problem faced by every data scientist: *how can you store and manipulate data without making errors?* However, this is not the only problem that you will face as a data scientist. The next problem will appear when you try to understand the information contained in your data. It is nearly impossible to spot insights or to discover patterns in raw data. A third problem will appear when you try to use your data set to reason about reality, which includes things not contained in your data set. What exactly does your data imply about things outside of the data set? How certain can you be?

I refer to these problems as the logistical, tactical, and strategic problems of data science, as shown in Figure 10-4. You'll face them whenever you try to learn from data:

A logistical problem
    How can you store and manipulate data without making errors?

A tactical problem
    How can you discover the information contained in your data?

A strategic problem
   How can you use the data to draw conclusions about the world at large?

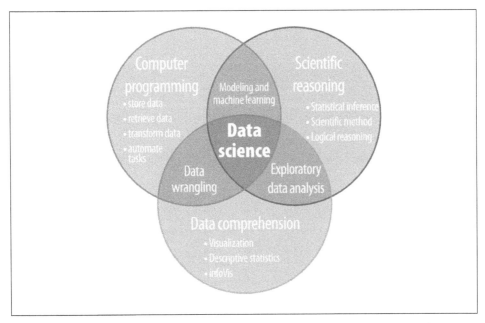

Figure 10-4. The three core skill sets of data science: computer programming, data comprehension, and scientific reasoning.

A well-rounded data scientist will need to be able to solve each of these problems in many different situations. By learning to program in R, you have mastered the logistical problem, which is a prerequisite for solving the tactical and strategic problems.

If you would like to learn how to reason with data, or how to transform, visualize, and explore your data sets with R tools, I recommend the book *Data Science with R*, the companion volume to this book. *Data Science with R* teaches a simple workflow for transforming, visualizing, and modeling data in R, as well as how to report results with the R Markdown and Shiny packages. More importantly, *Data Science with R* will teach you how to use your data to draw conclusions about the world at large, which is the real science of data science.

APPENDIX A
Installing R and RStudio

To get started with R, you need to acquire your own copy. This appendix will show you how to download R as well as RStudio, a software application that makes R easier to use. You'll go from downloading R to opening your first R session.

Both R and RStudio are free and easy to download.

How to Download and Install R

R is maintained by an international team of developers who make the language available through the web page of The Comprehensive R Archive Network (*http://cran.r-project.org*). The top of the web page provides three links for downloading R. Follow the link that describes your operating system: Windows, Mac, or Linux.

Windows

To install R on Windows, click the "Download R for Windows" link. Then click the "base" link. Next, click the first link at the top of the new page. This link should say something like "Download R 3.0.3 for Windows," except the 3.0.3 will be replaced by the most current version of R. The link downloads an installer program, which installs the most up-to-date version of R for Windows. Run this program and step through the installation wizard that appears. The wizard will install R into your program files folders and place a shortcut in your Start menu. Note that you'll need to have all of the appropriate administration privileges to install new software on your machine.

Mac

To install R on a Mac, click the "Download R for Mac" link. Next, click on the R-3.0.3 package link (or the package link for the most current release of R). An installer will download to guide you through the installation process, which is very easy. The installer lets you customize your installation, but the defaults will be suitable for most users. I've never found a reason to change them. If your computer requires a password before installing new progams, you'll need it here.

Binaries Versus Source

R can be installed from precompiled binaries or built from source on any operating system. For Windows and Mac machines, installing R from binaries is extremely easy. The binary comes preloaded in its own installer. Although you can build R from source on these platforms, the process is much more complicated and won't provide much benefit for most users. For Linux systems, the opposite is true. Precompiled binaries can be found for some systems, but it is much more common to build R from source files when installing on Linux. The download pages on CRAN's website (*http://cran.r-project.org*) provide information about building R from source for the Windows, Mac, and Linux platforms.

Linux

R comes preinstalled on many Linux systems, but you'll want the newest version of R if yours is out of date. The CRAN website (*http://cran.r-project.org*) provides files to build R from source on Debian, Redhat, SUSE, and Ubuntu systems under the link "Download R for Linux." Click the link and then follow the directory trail to the version of Linux you wish to install on. The exact installation procedure will vary depending on the Linux system you use. CRAN guides the process by grouping each set of source files with documentation or README files that explain how to install on your system.

32-bit Versus 64-bit

R comes in both 32-bit and 64-bit versions. Which should you use? In most cases, it won't matter. Both versions use 32-bit integers, which means they compute numbers to the same numerical precision. The difference occurs in the way each version manages memory. 64-bit R uses 64-bit memory pointers, and 32-bit R uses 32-bit memory pointers. This means 64-bit R has a larger memory space to use (and search through).

As a rule of thumb, 32-bit builds of R are faster than 64-bit builds, though not always. On the other hand, 64-bit builds can handle larger files and data sets with fewer memory management problems. In either version, the maximum allowable vector size tops out at around 2 billion elements. If your operating system doesn't support 64-bit programs, or your RAM is less than 4 GB, 32-bit R is for you. The Windows and Mac installers will automatically install both versions if your system supports 64-bit R.

Using R

R isn't a program that you can open and start using, like Microsoft Word or Internet Explorer. Instead, R is a computer language, like C, C++, or UNIX. You use R by writing commands in the R language and asking your computer to interpret them. In the old days, people ran R code in a UNIX terminal window—as if they were hackers in a movie from the 1980s. Now almost everyone uses R with an application called RStudio, and I recommend that you do, too.

R and UNIX

You can still run R in a UNIX or BASH window by typing the command:

R

which opens an R interpreter. You can then do your work and close the interpreter by running **q()** when you are finished.

RStudio

RStudio *is* an application like Microsoft Word—except that instead of helping you write in English, RStudio helps you write in R. I use RStudio throughout the book because it makes using R much easier. Also, the RStudio interface looks the same for Windows, Mac OS, and Linux. That will help me match the book to your personal experience.

You can download RStudio (*http://www.rstudio.com/ide*) for free. Just click the "Download RStudio" button and follow the simple instructions that follow. Once you've installed RStudio, you can open it like any other program on your computer—usually by clicking an icon on your desktop.

The R GUIs

Windows and Mac users usually do not program from a terminal window, so the Windows and Mac downloads for R come with a simple program that opens a terminal-like window for you to run R code in. This is what opens when you click the R icon on your Windows or Mac computer. These programs do a little more than the basic terminal window, but not much. You may hear people refer to them as the Windows or Mac R GUIs.

When you open RStudio, a window appears with three panes in it, as in Figure A-1. The largest pane is a console window. This is where you'll run your R code and see results. The console window is exactly what you'd see if you ran R from a UNIX console or the Windows or Mac GUIs. Everything else you see is unique to RStudio. Hidden in the other panes are a text editor, a graphics window, a debugger, a file manager, and much more. You'll learn about these panes as they become useful throughout the course of this book.

Figure A-1. The RStudio IDE for R.

Do I still need to download R?
Even if you use RStudio, you'll still need to download R to your computer. RStudio helps you use the version of R that lives on your computer, but it doesn't come with a version of R on its own.

Opening R

Now that you have both R and RStudio on your computer, you can begin using R by opening the RStudio program. Open RStudio just as you would any program, by clicking on its icon or by typing "RStudio" at the Windows Run prompt.

APPENDIX B
R Packages

Many of R's most useful functions do not come preloaded when you start R, but reside in *packages* that can be installed on top of R. R packages are similar to libraries in C, C++, and Javascript, packages in Python, and gems in Ruby. An R package bundles together useful functions, help files, and data sets. You can use these functions within your own R code once you load the package they live in. Usually the contents of an R package are all related to a single type of task, which the package helps solve. R packages will let you take advantage of R's most useful features: its large community of package writers (many of whom are active data scientists) and its prewritten routines for handling many common (and exotic) data-science tasks.

Base R
You may hear R users (or me) refer to "base R." What is base R? It is just the collection of R functions that gets loaded every time you start R. These functions provide the basics of the language, and you don't have to load a package before you can use them.

Installing Packages

To use an R package, you must first install it on your computer and then load it in your current R session. The easiest way to install an R package is with the install.packages R function. Open R and type the following into the command line:

```
install.packages("package name")
```

This will search for the specified package in the collection of packages hosted on the CRAN site. When R finds the package, it will download it into a libraries folder on your computer. R can access the package here in future R sessions without reinstalling it. Anyone can write an R package and disseminate it as they like; however, almost all R packages are published through the CRAN website. CRAN tests each R package before

193

publishing it. This doesn't eliminate every bug inside a package, but it does mean that you can trust a package on CRAN to run in the current version of R on your OS.

You can install multiple packages at once by linking their names with R's concatenate function, c. For example, to install the ggplot2, reshape2, and dplyr packages, run:

```
install.packages(c("ggplot2", "reshape2", "dplyr"))
```

If this is your first time installing a package, R will prompt you to choose an online mirror to install from. Mirrors are listed by location. Your downloads should be quickest if you select a mirror that is close to you. If you want to download a new package, try the Austria mirror first. This is the main CRAN repository, and new packages can sometimes take a couple of days to make it around to all of the other mirrors.

Loading Packages

Installing a package doesn't immediately place its functions at your fingertips. It just places them on your computer. To use an R package, you next have to load it in your R session with the command:

```
library(package name)
```

Notice that the quotation marks have disappeared. You can use them if you like, but quotation marks are optional for the library command. (This is not true for the install.packages command).

library will make all of the package's functions, data sets, and help files available to you until you close your current R session. The next time you begin an R session, you'll have to reload the package with library if you want to use it, but you won't have to reinstall it. You only have to install each package once. After that, a copy of the package will live in your R library. To see which packages you currently have in your R library, run:

```
library()
```

library() also shows the path to your actual R library, which is the folder that contains your R packages. You may notice many packages that you don't remember installing. This is because R automatically downloads a set of useful packages when you first install R.

Install packages from (almost) anywhere

The devtools R package makes it easy to install packages from locations other than the CRAN website. devtools provides functions like install_github, install_gitorious, install_bitbucket, and install_url. These work similar to install.packages, but they search new locations for R packages. install_github is especially useful because many R developers provide development versions of their packages on GitHub. The development version of a package will contain a sneak peek of new functions and patches but may not be as stable or as bug free as the CRAN version.

Why does R make you bother with installing and loading packages? You can imagine an R where every package came preloaded, but this would be a very large and slow program. As of May 6, 2014, the CRAN website hosts 5,511 packages. It is simpler to only install and load the packages that you want to use when you want to use them. This keeps your copy of R fast because it has fewer functions and help pages to search through at any one time. The arrangement has other benefits as well. For example, it is possible to update your copy of an R package without updating your entire copy of R.

What's the best way to learn about R packages?

It is difficult to use an R package if you don't know that it exists. You could go to the CRAN website and click the Packages link to see a list of available packages, but you'll have to wade through thousands of them. Moreover, many R packages do the same things.

How do you know which package does them best? The R-packages mailing list (*http://stat.ethz.ch/mailman/listinfo/r-packages*) is a place to start. It sends out announcements of new packages and maintains an archive of old announcements. Blogs that aggregate posts about R can also provide valuable leads. I recommend www.r-bloggers.com[R-bloggers]. RStudio maintains a list of some of the most useful R packages in the Getting Started section of *http://support.rstudio.com*. Finally, CRAN groups together some of the most useful—and most respected—packages by subject area (*http://cran.r-project.org/web/views*). This is an excellent place to learn about the packages designed for your area of work.

APPENDIX C
Updating R and Its Packages

The R Core Development Team continuously hones the R language by catching bugs, improving performance, and updating R to work with new technologies. As a result, new versions of R are released several times a year. The easiest way to stay current with R is to periodically check the CRAN website (*http://cran.r-project.org*). The website is updated for each new release and makes the release available for download. You'll have to install the new release. The process is the same as when you first installed R.

Don't worry if you're not interested in staying up-to-date on R Core's doings. R changes only slightly between releases, and you're not likely to notice the differences. However, updating to the current version of R is a good place to start if you ever encounter a bug that you can't explain.

RStudio also constantly improves its product. You can acquire the newest updates just by downloading them from RStudio (*http://www.rstudio.com/ide*).

R Packages

Package authors occasionally release new versions of their packages to add functions, fix bugs, or improve performance. The `update.packages` command checks whether you have the most current version of a package and installs the most current version if you do not. The syntax for `update.packages` follows that of `install.packages`. If you already have ggplot2, reshape2, and dplyr on your computer, it'd be a good idea to check for updates before you use them:

```
update.packages(c("ggplot2", "reshape2", "dplyr"))
```

You should start a new R session after updating packages. If you have a package loaded when you update it, you'll have to close your R session and open a new one to begin using the updated version of the package.

APPENDIX D
Loading and Saving Data in R

This appendix will show you how to load and save data into R from plain-text files, R files, and Excel spreadsheets. It will also show you the R packages that you can use to load data from databases and other common programs, like SAS and MATLAB.

Data Sets in Base R

R comes with many data sets preloaded in the `datasets` package, which comes with base R. These data sets are not very interesting, but they give you a chance to test code or make a point without having to load a data set from outside R. You can see a list of R's data sets as well as a short description of each by running:

```
help(package = "datasets")
```

To use a data set, just type its name. Each data set is already presaved as an R object. For example:

```
iris
##   Sepal.Length Sepal.Width Petal.Length Petal.Width Species
## 1          5.1         3.5          1.4         0.2  setosa
## 2          4.9         3.0          1.4         0.2  setosa
## 3          4.7         3.2          1.3         0.2  setosa
## 4          4.6         3.1          1.5         0.2  setosa
## 5          5.0         3.6          1.4         0.2  setosa
## 6          5.4         3.9          1.7         0.4  setosa
```

However, R's data sets are no substitute for your own data, which you can load into R from a wide variety of file formats. But before you load any data files into R, you'll need to determine where your *working directory* is.

Working Directory

Each time you open R, it links itself to a directory on your computer, which R calls the working directory. This is where R will look for files when you attempt to load them, and it is where R will save files when you save them. The location of your working directory will vary on different computers. To determine which directory R is using as your working directory, run:

```
getwd()
## "/Users/garrettgrolemund"
```

You can place data files straight into the folder that is your working directory, or you can move your working directory to where your data files are. You can move your working directory to any folder on your computer with the function `setwd`. Just give `setwd` the file path to your new working directory. I prefer to set my working directory to a folder dedicated to whichever project I am currently working on. That way I can keep all of my data, scripts, graphs, and reports in the same place. For example:

```
setwd("~/Users/garrettgrolemund/Documents/Book_Project")
```

If the file path does not begin with your root directory, R will assume that it begins at your current working directory.

You can also change your working directory by clicking on Session > Set Working Directory > Choose Directory in the RStudio menu bar. The Windows and Mac GUIs have similar options. If you start R from a UNIX command line (as on Linux machines), the working directory will be whichever directory you were in when you called R.

You can see what files are in your working directory with `list.files()`. If you see the file that you would like to open in your working directory, then you are ready to proceed. How you open files in your working directory will depend on which type of file you would like to open.

Plain-text Files

Plain-text files are one of the most common ways to save data. They are very simple and can be read by many different computer programs—even the most basic text editors. For this reason, public data often comes as plain-text files. For example, the Census Bureau, the Social Security Administration, and the Bureau of Labor Statistics all make their data available as plain-text files.

Here's how the royal flush data set from Chapter 3 would appear as a plain-text file (I've added a value column):

```
"card", "suit", "value"
"ace", "spades", 14
"king", "spades", 13
"queen", "spades", 12
```

```
"jack", "spades", 11
"ten", "spades", 10
```

A plain-text file stores a table of data in a text document. Each row of the table is saved on its own line, and a simple convention is used to separate the cells within a row. Often cells are separated by a comma, but they can also be separated by a tab, a pipe delimiter (i.e., |), or any other character. Each file only uses one method of separating cells, which minimizes confusion. Within each cell, data appears as you'd expect to see it, as words and numbers.

All plain-text files can be saved with the extension *.txt* (for text), but sometimes a file will receive a special extension that advertises how it separates data-cell entries. Since entries in the data set mentioned earlier are separated with a comma, this file would be a *comma-separated-values* file and would usually be saved with the extension *.csv*.

read.table

To load a plain-text file, use `read.table`. The first argument of `read.table` should be the name of your file (if it is in your working directory), or the file path to your file (if it is not in your working directory). If the file path does not begin with your root directory, R will append it to the end of the file path that leads to your working directory. You can give `read.table` other arguments as well. The two most important are `sep` and `header`.

If the royal flush data set was saved as a file named *poker.csv* in your working directory, you could load it with:

```
poker <- read.table("poker.csv", sep = ",", header = TRUE)
```

sep

Use `sep` to tell `read.table` what character your file uses to separate data entries. To find this out, you might have to open your file in a text editor and look at it. If you don't specify a `sep` argument, `read.table` will try to separate cells whenever it comes to white space, such as a tab or space. R won't be able to tell you if `read.table` does this correctly or not, so rely on it at your own risk.

header

Use `header` to tell `read.table` whether the first line of the file contains variable names instead of values. If the first line of the file is a set of variable names, you should set `header = TRUE`.

na.strings

Oftentimes data sets will use special symbols to represent missing information. If you know that your data uses a certain symbol to represent missing entries, you can tell `read.table` (and the preceding functions) what the symbol is with the `na.strings` argument. `read.table` will convert all instances of the missing information symbol to NA, which is R's missing information symbol (see "Missing Information" on page 89).

For example, your poker data set contained missing values stored as a `.`, like this:

```
## "card","suit","value"
## "ace"," spades"," 14"
## "king"," spades"," 13"
## "queen",".","."
## "jack",".","."
## "ten",".","."
```

You could read the data set into R and convert the missing values into NAs as you go with the command:

```
poker <- read.table("poker.csv", sep = ",", header = TRUE, na.string = ".")
```

R would save a version of poker that looks like this:

```
##   card   suit value
##    ace spades    14
##   king spades    13
##  queen   <NA>    NA
##   jack   <NA>    NA
##    ten   <NA>    NA
```

skip and nrow

Sometimes a plain-text file will come with introductory text that is not part of the data set. Or, you may decide that you only wish to read in part of a data set. You can do these things with the `skip` and `nrow` arguments. Use `skip` to tell R to skip a specific number of lines before it starts reading in values from the file. Use `nrow` to tell R to stop reading in values after it has read in a certain number of lines.

For example, imagine that the complete royal flush file looks like this:

```
This data was collected by the National Poker Institute.
We accidentally repeated the last row of data.

"card", "suit", "value"
"ace", "spades", 14
"king", "spades", 13
"queen", "spades", 12
"jack", "spades", 11
"ten", "spades", 10
"ten", "spades", 10
```

You can read just the six lines (five rows plus a header) that you want with:

```
read.table("poker.csv", sep = ",", header = TRUE, skip = 3, nrow = 5)
##   card  suit value
## 1  ace spades    14
```

```
## 2  king  spades  13
## 3 queen  spades  12
## 4  jack  spades  11
## 5   ten  spades  10
```

Notice that the header row doesn't count towards the total rows allowed by nrow.

stringsAsFactors

R reads in numbers just as you'd expect, but when R comes across character strings (e.g., letters and words) it begins to act strangely. R wants to convert every character string into a factor. This is R's default behavior, but I think it is a mistake. Sometimes factors are useful. At other times, they're clearly the wrong data type for the job. Also factors cause weird behavior, especially when you want to display data. This behavior can be surprising if you didn't realize that R converted your data to factors. In general, you'll have a smoother R experience if you don't let R make factors until you ask for them. Thankfully, it is easy to do this.

Setting the argument stringsAsFactors to FALSE will ensure that R saves any character strings in your data set as character strings, not factors. To use stringsAsFactors, you'd write:

```
read.table("poker.csv", sep = ",", header = TRUE, stringsAsFactors = FALSE)
```

If you will be loading more than one data file, you can change the default factoring behavior at the global level with:

```
options(stringsAsFactors = FALSE)
```

This will ensure that all strings will be read as strings, not as factors, until you end your R session, or rechange the global default by running:

```
options(stringsAsFactors = TRUE)
```

The read Family

R also comes with some prepackaged short cuts for read.table, shown in Table D-1.

Table D-1. R's read functions. You can overwrite any of the default arguments as necessary.

Function	Defaults	Use
read.table	sep = " ", header = FALSE	General-purpose read function
read.csv	sep = ",", header = TRUE	Comma-separated-variable (CSV) files
read.delim	sep = "\t", header = TRUE	Tab-delimited files
read.csv2	sep = ";", header = TRUE, dec = ","	CSV files with European decimal format
read.delim2	sep = "\t", header = TRUE, dec = ","	Tab-delimited files with European decimal format

The first shortcut, `read.csv`, behaves just like `read.table` but automatically sets `sep = ","` and `header = TRUE`, which can save you some typing:

```
poker <- read.csv("poker.csv")
```

`read.delim` automatically sets `sep` to the tab character, which is very handy for reading tab delimited files. These are files where each cell is separated by a tab. `read.delim` also sets `header = TRUE` by default.

`read.delim2` and `read.csv2` exist for European R users. These functions tell R that the data uses a comma instead of a period to denote decimal places. (If you're wondering how this works with CSV files, CSV2 files usually separate cells with a semicolon, not a comma.)

Import Dataset
You can also load plain text files with RStudio's Import Dataset button, as described in "Loading Data" on page 57. Import Dataset provides a GUI version of `read.table`.

read.fwf

One type of plain-text file defies the pattern by using its layout to separate data cells. Each row is placed in its own line (as with other plain-text files), and then each column begins at a specific number of characters from the lefthand side of the document. To achieve this, an arbitrary number of character spaces is added to the end of each entry to correctly position the next entry. These documents are known as *fixed-width files* and usually end with the extension *.fwf*.

Here's one way the royal flush data set could look as a fixed-width file. In each row, the suit entry begins exactly 10 characters from the start of the line. It doesn't matter how many characters appeared in the first cell of each row:

```
card      suit      value
ace       spades    14
king      spades    13
queen     spades    12
jack      spades    11
10        spades    10
```

Fixed-width files look nice to human eyes (but no better than a tab-delimited file); however, they can be difficult to work with. Perhaps because of this, R comes with a function for reading fixed-width files, but no function for saving them. Unfortunately, US government agencies seem to like fixed-width files, and you'll likely encounter one or more during your career.

You can read fixed-width files into R with the function `read.fwf`. The function takes the same arguments as `read.table` but requires an additional argument, `widths`, which should be a vector of numbers. Each *i*th entry of the `widths` vector should state the width (in characters) of the *i*th column of the data set.

If the aforementioned fixed-width royal flush data was saved as *poker.fwf* in your working directory, you could read it with:

```
poker <- read.fwf("poker.fwf", widths = c(10, 7, 6), header = TRUE)
```

HTML Links

Many data files are made available on the Internet at their own web address. If you are connected to the Internet, you can open these files straight into R with `read.table`, `read.csv`, etc. You can pass a web address into the file name argument for any of R's data-reading functions. As a result, you could read in the poker data set from a web address like *http://.../poker.csv* with:

```
poker <- read.csv("http://.../poker.csv")
```

Just make sure that the web address links directly to the file and not to a web page that links to the file. Usually, when you visit a data file's web address, the file will begin to download or the raw data will appear in your browser window.

Note that websites that begin with *https://* are secure websites, which means R may not be able to access the data provided at these links.

Saving Plain-Text Files

Once your data is in R, you can save it to any file format that R supports. If you'd like to save it as a plain-text file, you can use the `write` family of functions. The three basic write functions appear in Table D-2. Use `write.csv` to save your data as a *.csv* file and `write.table` to save your data as a tab delimited document or a document with more exotic separators.

Table D-2. R saves data sets to plain-text files with the write family of functions

File format	Function and syntax
.csv	`write.csv(r_object, file = filepath, row.names = FALSE)`
.csv (with European decimal notation)	`write.csv2(r_object, file = filepath, row.names = FALSE)`
tab delimited	`write.table(r_object, file = filepath, sep = "\t", row.names=FALSE)`

The first argument of each function is the R object that contains your data set. The `file` argument is the file name (including extension) that you wish to give the saved data. By default, each function will save your data into your working directory. However, you can supply a file path to the file argument. R will oblige by saving the file at the end of the file path. If the file path does not begin with your root directory, R will append it to the end of the file path that leads to your working directory.

For example, you can save the (hypothetical) poker data frame to a subdirectory named *data* within your working directory with the command:

```
write.csv(poker, "data/poker.csv", row.names = FALSE)
```

Keep in mind that `write.csv` and `write.table` cannot create new directories on your computer. Each folder in the file path must exist before you try to save a file with it.

The `row.names` argument prevents R from saving the data frame's row names as a column in the plain-text file. You might have noticed that R automatically names each row in a data frame with a number. For example, each row in our poker data frame appears with a number next to it:

```
poker
##     card  suit value
## 1    ace spades    14
## 2   king spades    13
## 3  queen spades    12
## 4   jack spades    11
## 5     10 spades    10
```

These row numbers are helpful, but can quickly accumulate if you start saving them. R will add a new set of numbers by default each time you read the file back in. Avoid this by always setting `row.names = FALSE` when you use a function in the `write` family.

Compressing Files

To compress a plain-text file, surround the file name or file path with the function `bzfile`, `gzfile`, or `xzfile`. For example:

```
write.csv(poker, file = bzfile("data/poker.csv.bz2"), row.names = FALSE)
```

Each of these functions will compress the output with a different type of compression format, shown in Table D-3.

Table D-3. R comes with three helper functions for compressing files

Function	Compression type
bzfile	bzip2
gzfile	gnu zip (gzip)
xzfile	xz compression

It is a good idea to adjust your file's extension to reflect the compression. R's read functions will open plain-text files compressed in any of these formats. For example, you could read a compressed file named *poker.csv.bz2* with:

```
read.csv("poker.csv.bz2")
```

or:

```
read.csv("data/poker.csv.bz2")
```

depending on where the file is saved.

R Files

R provides two file formats of its own for storing data, *.RDS* and *.RData*. RDS files can store a single R object, and RData files can store multiple R objects.

You can open a RDS file with readRDS. For example, if the royal flush data was saved as *poker.RDS*, you could open it with:

```
poker <- readRDS("poker.RDS")
```

Opening RData files is even easier. Simply run the function load with the file:

```
load("file.RData")
```

There's no need to assign the output to an object. The R objects in your RData file will be loaded into your R session with their original names. RData files can contain multiple R objects, so loading one may read in multiple objects. load doesn't tell you how many objects it is reading in, nor what their names are, so it pays to know a little about the RData file before you load it.

If worse comes to worst, you can keep an eye on the environment pane in RStudio as you load an RData file. It displays all of the objects that you have created or loaded during your R session. Another useful trick is to put parentheses around your load command like so, (load("poker.RData")). This will cause R to print out the names of each object it loads from the file.

Both readRDS and load take a file path as their first argument, just like R's other read and write functions. If your file is in your working directory, the file path will be the file name.

Saving R Files

You can save an R object like a data frame as either an RData file or an RDS file. RData files can store multiple R objects at once, but RDS files are the better choice because they foster reproducible code.

To save data as an RData object, use the save function. To save data as a RDS object, use the saveRDS function. In each case, the first argument should be the name of the R object you wish to save. You should then include a file argument that has the file name or file path you want to save the data set to.

For example, if you have three R objects, a, b, and c, you could save them all in the same RData file and then reload them in another R session:

```
a <- 1
b <- 2
c <- 3
save(a, b, c, file = "stuff.RData")
load("stuff.RData")
```

However, if you forget the names of your objects or give your file to someone else to use, it will be difficult to determine what was in the file—even after you (or they) load it. The user interface for RDS files is much more clear. You can save only one object per file, and whoever loads it can decide what they want to call their new data. As a bonus, you don't have to worry about load overwriting any R objects that happened to have the same name as the objects you are loading:

```
saveRDS(a, file = "stuff.RDS")
a <- readRDS("stuff.RDS")
```

Saving your data as an R file offers some advantages over saving your data as a plaintext file. R automatically compresses the file and will also save any R-related metadata associated with your object. This can be handy if your data contains factors, dates and times, or class attributes. You won't have to reparse this information into R the way you would if you converted everything to a text file.

On the other hand, R files cannot be read by many other programs, which makes them inefficient for sharing. They may also create a problem for long-term storage if you don't think you'll have a copy of R when you reopen the files.

Excel Spreadsheets

Microsoft Excel is a popular spreadsheet program that has become almost industry standard in the business world. There is a good chance that you will need to work with an Excel spreadsheet in R at least once in your career. You can read spreadsheets into R and also save R data as a spreadsheet in a variety of ways.

Export from Excel

The best method for moving data from Excel to R is to export the spreadsheet from Excel as a *.csv* or *.txt* file. Not only will R be able to read the text file, so will any other data analysis software. Text files are the lingua franca of data storage.

Exporting the data solves another difficulty as well. Excel uses proprietary formats and metadata that will not easily transfer into R. For example, a single Excel file can include multiple spreadsheets, each with their own columns and macros. When Excel exports the file as a *.csv* or *.txt*, it makes sure this format is transferred into a plain-text file in the most appropriate way. R may not be able to manage the conversion as efficiently.

To export data from Excel, open the Excel spreadsheet and then go to Save As in the Microsoft Office Button menu. Then choose CSV in the Save as type box that appears and save the files. You can then read the file into R with the `read.csv` function.

Copy and Paste

You can also copy portions of an Excel spreadsheet and paste them into R. To do this, open the spreadsheet and select the cells you wish to read into R. Then select Edit > Copy in the menu bar—or use a keyboard shortcut—to copy the cells to your clipboard.

On most operating systems, you can read the data stored in your clipboard into R with:

```
read.table("clipboard")
```

On Macs you will need to use:

```
read.table(pipe("pbpaste"))
```

If the cells contain values with spaces in them, this will disrupt `read.table`. You can try another `read` function (or just formally export the data from Excel) before reading it into R.

XLConnect

Many packages have been written to help you read Excel files directly into R. Unfortunately, many of these packages do not work on all operating systems. Others have been made out of date by the *.xlsx* file format. One package that does work on all file systems (and gets good reviews) is the XLConnect package. To use it, you'll need to install and load the package:

```
install.packages("XLConnect")
library(XLConnect)
```

XLConnect relies on Java to be platform independent. So when you first open XLConnect, RStudio may ask to download a Java Runtime Environment if you do not already have one.

Reading Spreadsheets

You can use XLConnect to read in an Excel spreadsheet with either a one- or a two-step process. I'll start with the two-step process. First, load an Excel workbook with `loadWorkbook`. `loadWorkbook` can load both *.xls* and *.xlsx* files. It takes one argument: the file path to your Excel workbook (this will be the name of the workbook if it is saved in your working directory):

```
wb <- loadWorkbook("file.xlsx")
```

Next, read a spreadsheet from the workbook with `readWorksheet`, which takes several arguments. The first argument should be a workbook object created with `loadWork` book. The next argument, sheet, should be the name of the spreadsheet in the workbook that you would like to read into R. This will be the name that appears on the bottom tab of the spreadsheet. You can also give sheet a number, which specifies the sheet that you want to read in (one for the first sheet, two for the second, and so on).

readWorksheet then takes four arguments that specify a bounding box of cells to read in: startRow, startCol, endRow, and endCol. Use startRow and startCol to describe the cell in the top-left corner of the bounding box of cells that you wish to read in. Use endRow and endCol to specify the cell in the bottom-right corner of the bounding box. Each of these arguments takes a number. If you do not supply bounding arguments, readWorksheet will read in the rectangular region of cells in the spreadsheet that appears to contain data. readWorksheet will assume that this region contains a header row, but you can tell it otherwise with header = FALSE.

So to read in the first worksheet from wb, you could use:

```
sheet1 <- readWorksheet(wb, sheet = 1, startRow = 0, startCol = 0,
    endRow = 100, endCol = 3)
```

R will save the output as a data frame. All of the arguments in readWorkbook except the first are vectorized, so you can use it to read in multiple sheets from the same workbook at once (or multiple cell regions from a single worksheet). In this case, readWork sheet will return a list of data frames.

You can combine these two steps with readWorksheetFromFile. It takes the file argument from loadWorkbook and combines it with the arguments from readWorksheet. You can use it to read one or more sheets straight from an Excel file:

```
sheet1 <- readWorksheetFromFile("file.xlsx", sheet = 1, startRow = 0,
    startCol = 0, endRow = 100, endCol = 3)
```

Writing Spreadsheets

Writing to an Excel spreadsheet is a four-step process. First, you need to set up a workbook object with loadWorkbook. This works just as before, except if you are not using an existing Excel file, you should add the argument create = TRUE. XLConnect will create a blank workbook. When you save it, XLConnect will write it to the file location that you specified here with loadWorkbook:

```
wb <- loadWorkbook("file.xlsx", create = TRUE)
```

Next, you need to create a worksheet inside your workbook object with createSheet. Tell createSheet which workbook to place the sheet in and which to use for the sheet.

```
createSheet(wb, "Sheet 1")
```

Then you can save your data frame or matrix to the sheet with writeWorksheet. The first argument of writeWorksheet, object, is the workbook to write the data to. The second argument, `data`, is the data to write. The third argument, `sheet`, is the name of the sheet to write it to. The next two arguments, `startRow` and `startCol`, tell R where in the spreadsheet to place the upper-left cell of the new data. These arguments each default to 1. Finally, you can use `header` to tell R whether your column names should be written with the data:

```
writeWorksheet(wb, data = poker, sheet = "Sheet 1")
```

Once you have finished adding sheets and data to your workbook, you can save it by running `saveWorkbook` on the workbook object. R will save the workbook to the file name or path you provided in `loadWorkbook`. If this leads to an existing Excel file, R will overwrite it. If it leads to a new file, R will create it.

You can also collapse these steps into a single call with `writeWorksheetToFile`, like this:

```
writeWorksheetToFile("file.xlsx", data = poker, sheet = "Sheet 1",
    startRow = 1, startCol = 1)
```

The XLConnect package also lets you do more advanced things with Excel spreadsheets, such as writing to a named region in a spreadsheet, working with formulas, and assigning styles to cells. You can read about these features in XLConnect's vignette, which is accessible by loading XLConnect and then running:

```
vignette("XLConnect")
```

Loading Files from Other Programs

You should follow the same advice I gave you for Excel files whenever you wish to work with file formats native to other programs: open the file in the original program and export the data as a plain-text file, usually a CSV. This will ensure the most faithful transcription of the data in the file, and it will usually give you the most options for customizing how the data is transcribed.

Sometimes, however, you may acquire a file but not the program it came from. As a result, you won't be able to open the file in its native program and export it as a text file. In this case, you can use one of the functions in Table D-4 to open the file. These functions mostly come in R's `foreign` package. Each attempts to read in a different file format with as few hiccups as possible.

Table D-4. *A number of functions will attempt to read the file types of other data-analysis programs*

File format	Function	Library
ERSI ArcGIS	`read.shapefile`	shapefiles
Matlab	`readMat`	R.matlab
minitab	`read.mtp`	foreign
SAS (permanent data set)	`read.ssd`	foreign

File format	Function	Library
SAS (XPORT format)	`read.xport`	foreign
SPSS	`read.spss`	foreign
Stata	`read.dta`	foreign
Systat	`read.systat`	foreign

Connecting to Databases

You can also use R to connect to a database and read in data. How you do this will depend on the database management system that you use. Working with a database will require experience that goes beyond the skill set of a typical R user. However, if you are interested in doing this, the best place to start is by downloading these R packages and reading their documentation.

Use the RODBC package to connect to databases through an ODBC connection.

Use the DBI package to connect to databases through individual drivers. The DBI package provides a common syntax for working with different databases. You will have to download a database-specific package to use in conjunction with DBI. These packages provide the API for the native drivers of different database programs. For MySQL use RMySQL, for SQLite use RSQLite, for Oracle use ROracle, for PostgreSQL use RPostgreSQL, and for databases that use drivers based on the Java Database Connectivity (JDBC) API use RJDBC. Once you have loaded the appropriate driver package, you can use the commands provided by DBI to access your database.

APPENDIX E
Debugging R Code

This appendix refers to environments, the topic of Chapter 6, and uses examples from Chapter 7 and Chapter 8. You should read through these chapters first to get the most out of this appendix.

R comes with a simple set of debugging tools that RStudio amplifies. You can use these tools to better understand code that produces an error or returns an unexpected result. Usually this will be your own code, but you can also examine the functions in R or one of its packages.

Debugging code can take as much creativity and insight as writing code. There is no guarantee that you will find a bug or be able to fix it when you do. However, you can help yourself by using R's debugging tools. These include the traceback, browser, debug, debugonce, trace, and recover functions.

Using these tools is usually a two-step process. First, you locate *where* an error occurred. Then you try to determine *why* it occurred. You can do the first step with R's trace back function.

traceback

The traceback tool pinpoints the location of an error. Many R functions call other R functions, which call other functions, and so on. When an error occurs, it may not be clear which of these functions went wrong. Let's consider an example. The following functions call one another, and the last function creates an error (you'll see why in a second):

```
first <- function() second()
second <- function() third()
```

```
third <- function() fourth()
fourth <- function() fifth()
fifth <- function() bug()
```

When you run `first`, it will call `second`, which will call `third`, which will call `fourth`, which will call `fifth`, which will call `bug`, a function that does not exist. Here's what that will look like at the command line:

```
first()
## Error in fifth() : could not find function "bug"
```

The error report tells us that the error occurred when R tried to run `fifth`. It also tells us the nature of the error (there is no function called `bug`). Here, it is obvious why R calls `fifth`, but it might not be so obvious why R calls a function when an error occurs in the wild.

You can see the path of functions that R called before it hit an error by typing **traceback()** at the command line. `traceback` will return a call stack, a list of the functions that R called in the order that it called them. The bottom function will be the command that you entered in the command line. The top function will be the function that caused the error:

```
traceback()
## 5: fifth() at #1
## 4: fourth() at #1
## 3: third() at #1
## 2: second() at #1
## 1: first()
```

`traceback` will always refer to the last error you encountered. If you would like to look at a less recent error, you will need to recreate it before running `traceback`.

How can this help you? First, `traceback` returns a list of suspects. One of these functions caused the error, and each function is more suspicious than the ones below it. Chances are that our bug came from `fifth` (it did), but it is also possible that an earlier function did something odd—like call `fifth` when it shouldn't have.

Second, `traceback` can show you if R stepped off the path that you expected it to take. If this happened, look at the last function before things went wrong.

Third, `traceback` can reveal the frightening extent of infinite recursion errors. For example, if you change `fifth` so that it calls `second`, the functions will make a loop: `second` will call `third`, which will call `fourth`, which will call `fifth`, which will call `second` and start the loop over again. It is easier to do this sort of thing in practice than you might think:

```
fifth <- function() second()
```

When you call `first()`, R will start to run the functions. After awhile, it will notice that it is repeating itself and will return an error. `traceback` will show just what R was doing:

```
first()
## Error: evaluation nested too deeply: infinite recursion/options(expressions=)?

traceback()
## 5000: fourth() at #1
## 4999: third() at #1
## 4998: second() at #1
## 4997: fifth() at #1
## 4996: fourth() at #1
## 4995: third() at #1
## 4994: second() at #1
## 4993: fifth() at #1
## ...
```

Notice that there are 5,000 lines of output in this `traceback`. If you are using RStudio, you will not get to see the traceback of an infinite recursion error (I used the Mac GUI to get this output). RStudio represses the traceback for infinite recursion errors to prevent the large call stacks from pushing your console history out of R's memory buffer. With RStudio, you will have to recognize the infinite recursion error by its error message. However, you can still see the imposing `traceback` by running things in a UNIX shell or the Windows or Mac GUIs.

RStudio makes it very easy to use `traceback`. You do not even need to type in the function name. Whenever an error occurs, RStudio will display it in a gray box with two options. The first is Show Traceback, shown in Figure E-1.

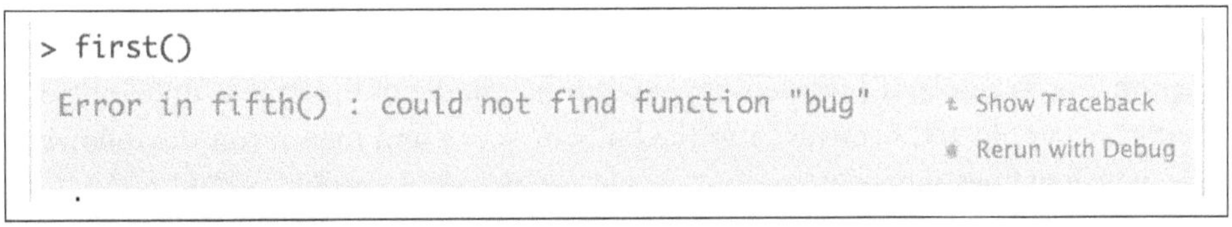

```
> first()
Error in fifth() : could not find function "bug"        ± Show Traceback
                                                        * Rerun with Debug
```

Figure E-1. RStudio's Show Traceback option.

If you click Show Traceback, RStudio will expand the gray box and display the trace back call stack, as in Figure E-2. The Show Traceback option will persist beside an error message in your console, even as you write new commands. This means that you can go back and look at the call stacks for all errors—not just the most recent error.

Imagine that you've used `traceback` to pinpoint a function that you think might cause a bug. Now what should you do? You should try to figure out what the function did to cause an error while it ran (if it did anything). You can examine how the function runs with `browser`.

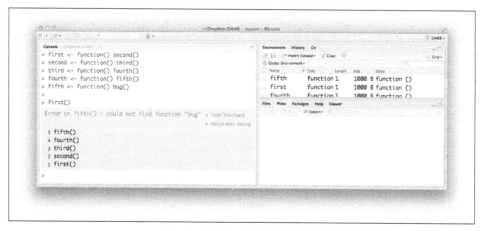

Figure E-2. RStudio's Traceback display.

browser

You can ask R to pause in the middle of running a function and give control back to you with browser. This will let you enter new commands at the command line. The active environment for these commands will not be the global environment (as usual); it will be the runtime environment of the function that you have paused. As a result, you can look at the objects that the function is using, look up their values with the same scoping rules that the function would use, and run code under the same conditions that the function would run it in. This arrangement provides the best chance for spotting the source of bugs in a function.

To use browser, add the call browser() to the body of a function and then resave the function. For example, if I wanted to pause in the middle of the score function from Chapter 7, I could add browser() to the body of score and then rerun the following code, which defines score:

```
score <- function (symbols) {
  # identify case
  same <- symbols[1] == symbols[2] && symbols[2] == symbols[3]
  bars <- symbols %in% c("B", "BB", "BBB")

  # get prize
  if (same) {
    payouts <- c("DD" = 100, "7" = 80, "BBB" = 40, "BB" = 25,
      "B" = 10, "C" = 10, "0" = 0)
    prize <- unname(payouts[symbols[1]])
  } else if (all(bars)) {
    prize <- 5
  } else {
    cherries <- sum(symbols == "C")
    prize <- c(0, 2, 5)[cherries + 1]
```

```
  }

  browser()

  # adjust for diamonds
  diamonds <- sum(symbols == "DD")
  prize * 2 ^ diamonds
}
```

Now whenever R runs `score`, it will come to the call `browser()`. You can see this with the `play` function from Chapter 7. If you don't have `play` handy, you can access it by running this code:

```
get_symbols <- function() {
  wheel <- c("DD", "7", "BBB", "BB", "B", "C", "0")
  sample(wheel, size = 3, replace = TRUE,
    prob = c(0.03, 0.03, 0.06, 0.1, 0.25, 0.01, 0.52))
}

play <- function() {
  symbols <- get_symbols()
  structure(score(symbols), symbols = symbols, class = "slots")
}
```

When you run `play`, `play` will call `get_symbols` and then `score`. As R works through `score`, it will come across the call to `browser` and run it. When R runs this call, several things will happen, as in Figure E-3. First, R will stop running `score`. Second, the command prompt will change to `browser[1]>` and R will give me back control; I can now type new commands in at the new command prompt. Third, three buttons will appear above the console pane: Next, Continue, and Stop. Fourth, RStudio will display the source code for `score` in the scripts pane, and it will highlight the line that contains `browser()`. Fifth, the environments tab will change. Instead of revealing the objects that are saved in the global environment, it will reveal the objects that are saved in the runtime environment of `score` (see Chapter 6 for an explanation of R's environment system). Sixth, RStudio will open a new Traceback pane, which shows the call stack RStudio took to get to `browser`. The most recent function, `score`, will be highlighted.

I'm now in a new R mode, called *browser mode*. Browser mode is designed to help you uncover bugs, and the new display in RStudio is designed to help you navigate this mode.

Any command that you run in browser mode will be evaluated in the context of the runtime environment of the function that called `browser`. This will be the function that is highlighted in the new Traceback pane. Here, that function is `score`. So while we are in browser mode, the active environment will be `score`'s runtime environment. This lets you do two things.

Figure E-3. RStudio updates its display whenever you enter browser mode to help you navigate the mode.

First, you can inspect the objects that score uses. The updated Environments pane shows you which objects score has saved in its local environment. You can inspect any of them by typing their name at the browser prompt. This gives you a way to see the values of runtime variables that you normally would not be able to access. If a value looks clearly wrong, you may be close to finding a bug:

```
Browse[1]> symbols
## [1] "B" "B" "0"

Browse[1]> same
## [1] FALSE
```

Second, you can run code and see the same results that score would see. For example, you could run the remaining lines of the score function and see if they do anything unusual. You could run these lines by typing them into the command prompt, or you could use the three navigation buttons that now appear above the prompt, as in Figure E-4.

The first button, Next, will run the next line of code in score. The highlighted line in the scripts pane will advance by one line to show you your new location in the score function. If the next line begins a code chunk, like a for loop or an if tree, R will run the whole chunk and will highlight the whole chunk in the script window.

The second button, Continue, will run all of the remaining lines of score and then exit the browser mode.

The third button, Stop, will exit browser mode without running any more lines of score.

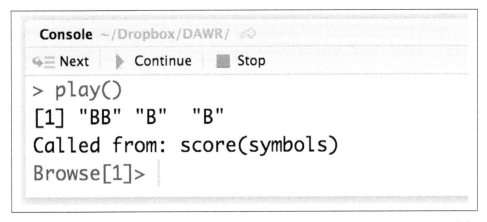

Figure E-4. You can navigate browser mode with the three buttons at the top of the console pane.

You can do the same things by typing the commands n, c, and Q into the browser prompt. This creates an annoyance: what if you want to look up an object named n, c, or Q? Typing in the object name will not work, R will either advance, continue, or quit the browser mode. Instead you will have to look these objects up with the commands get("n"), get("c"), and get("Q"). cont is a synonym for c in browser mode and where prints the call stack, so you'll have to look up these objects with get as well.

Browser mode can help you see things from the perspective of your functions, but it cannot show you where the bug lies. However, browser mode can help you test hypotheses and investigate function behavior. This is usually all you need to spot and fix a bug. The browser mode is the basic debugging tool of R. Each of the following functions just provides an alternate way to enter the browser mode.

Once you fix the bug, you should resave your function a third time—this time without the browser() call. As long as the browser call is in there, R will pause each time you, or another function, calls score.

Break Points

RStudio's break points provide a graphical way to add a browser statement to a function. To use them, open the script where you've defined a function. Then click to the left of the line number of the line of code in the function body where you'd like to add the browser statement. A hollow red dot will appear to show you where the break point will occur. Then run the script by clicking the Source button at the top of the Scripts pane. The hollow dot will turn into a solid red dot to show that the function has a break point (see Figure E-5).

R will treat the break point like a `browser` statement, going into browser mode when it encounters it. You can remove a break point by clicking on the red dot. The dot will disappear, and the break point will be removed.

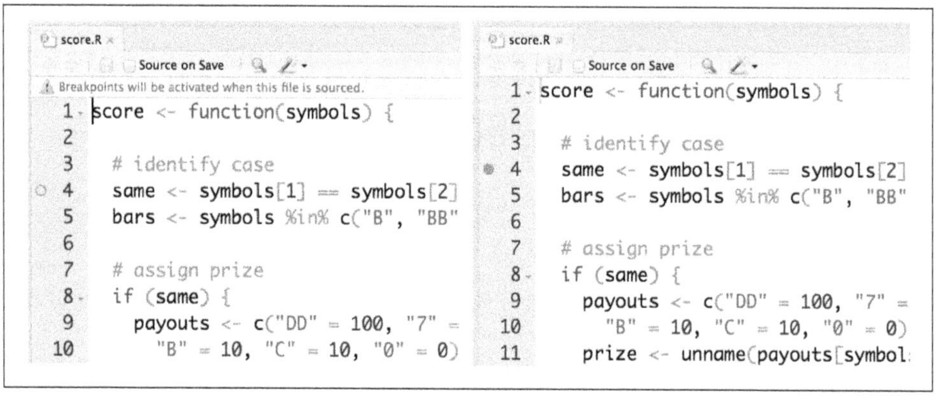

Figure E-5. Break points provide the graphical equivalent of a browser statement.

Break points and `browser` provide a great way to debug functions that you have defined. But what if you want to debug a function that already exists in R? You can do that with the `debug` function.

debug

You can "add" a browser call to the very start of a preexisting function with `debug`. To do this, run `debug` on the function. For example, you can run `debug` on `sample` with:

```
debug(sample)
```

Afterward, R will act as if there is a `browser()` statement in the first line of the function. Whenever R runs the function, it will immediately enter browser mode, allowing you to step through the function one line at a time. R will continue to behave this way until you "remove" the browser statement with `undebug`:

```
undebug(sample)
```

You can check whether a function is in "debugging" mode with `isdebugged`. This will return TRUE if you've ran `debug` on the function but have yet to run `undebug`:

```
isdebugged(sample)
## FALSE
```

If this is all too much of a hassle, you can do what I do and use `debugonce` instead of `debug`. R will enter browser mode the very next time it runs the function but will automatically undebug the function afterward. If you need to browse through the function again, you can just run debugonce on it a second time.

You can recreate debugonce in RStudio whenever an error occurs. "Rerun with debug" will appear in the grey error box beneath Show Traceback (Figure E-1). If you click this option, RStudio will rerun the command as if you had first run debugonce on it. R will immediately go into browser mode, allowing you to step through the code. The browser behavior will only occur on this run of the code. You do not need to worry about calling undebug when you are done.

trace

You can add the browser statement further into the function, and not at the very start, with trace. trace takes the name of a function as a character string and then an R expression to insert into the function. You can also provide an at argument that tells trace at which line of the function to place the expression. So to insert a browser call at the fourth line of sample, you would run:

```
trace("sample", browser, at = 4)
```

You can use trace to insert other R functions (not just browser) into a function, but you may need to think of a clever reason for doing so. You can also run trace on a function without inserting any new code. R will prints trace:<the function> at the command line every time R runs the function. This is a great way to test a claim I made in Chapter 8, that R calls print every time it displays something at the command line:

```
trace(print)

first
## trace: print(function () second())
## function() second()

head(deck)
## trace: print
##     face   suit value
## 1   king spades    13
## 2  queen spades    12
## 3   jack spades    11
## 4    ten spades    10
## 5   nine spades     9
## 6  eight spades     8
```

You can revert a function to normal after calling trace on it with untrace:

```
untrace(sample)
untrace(print)
```

recover

The `recover` function provides one final option for debugging. It combines the call stack of `traceback` with the browser mode of `browser`. You can use `recover` just like `browser`, by inserting it directly into a function's body. Let's demonstrate `recover` with the `fifth` function:

```
fifth <- function() recover()
```

When R runs `recover`, it will pause and display the call stack, but that's not all. R gives you the option of opening a browser mode in *any* of the functions that appear in the call stack. Annoyingly, the call stack will be displayed upside down compared to `trace back`. The most recent function will be on the bottom, and the original function will be on the top:

```
first()
##
## Enter a frame number, or 0 to exit
##
## 1: first()
## 2: #1: second()
## 3: #1: third()
## 4: #1: fourth()
## 5: #1: fifth()
```

To enter a browser mode, type in the number next to the function in whose runtime environment you would like to browse. If you do not wish to browse any of the functions, type 0:

```
3
## Selection: 3
## Called from: fourth()
## Browse[1]>
```

You can then proceed as normal. `recover` gives you a chance to inspect variables up and down your call stack and is a powerful tool for uncovering bugs. However, adding `recover` to the body of an R function can be cumbersome. Most R users use it as a global option for handling errors.

If you run the following code, R will automatically call `recover()` whenever an error occurs:

```
options(error = recover)
```

This behavior will last until you close your R session, or reverse the behavior by calling:

```
options(error = NULL)
```

Index

Symbols
!= operator, 81
" (quote mark), 41
(hashtag character), 6, 138
(double hashtag character), 6
$ (dollar sign), 73, 97
%*% operator, 11
%in% operator, 81
%o% operator, 11
& operator, 85, 130
&& operator, 130
) (parentheses), 26
+ operator, 150
+ prompt, 5
- operator, 150
.Call, 181
.Internal, 181
.Primitive, 181
: (colon operator), 5, 7
< operator, 81, 150
<- assignment operator, 77, 81, 97, 176
<= operator, 81
= (equals sign), 81
== operator, 81
> operator, 81
> prompt, 6
>= operator, 81
? (question mark), 29
[(hard bracket, single), 26, 65, 74

[1], 6
[[(hard brackets, double), 73
{} (braces), 17
| operator, 85, 130
|| operator, 130

A
accessor functions, 96
active environments, 97, 99
algebra, 66
all function, 85, 123
any function, 85, 123
args (), 14
arguments
    applying multiple, 13
    default values for, 19
    definition of, 12
    looking up, 14
    naming, 13, 18
arithmetic, basic, 6
array function, 46
as.character functions, 50
as.environment function, 95
assign function, 97
assignment, 99
assignment operator (<-), 77, 81, 97, 176
atomic vectors
    characters, 41
    class, 47

We'd like to hear your suggestions for improving our indexes. Send email to index@oreilly.com.

coercion in, 51
complex, 42
creating with c, 38
discovering type of, 39
doubles, 39, 48
integers, 40
logicals, 42
raw, 42
testing for, 38
types of, 38
attach(), 76
attr function, 143
attributes
    adding, 143
    as metadata, 43, 142
    class, 47
    dim, 45
    displaying, 43, 142
    looking up, 143
    names, 44
average value, 158

B

base R, 193, 199
binaries, installing R from, 190
binwidth argument, 26
blank space indexing, 69
boolean data, 42
boolean operators, 85
bounding arguments, 210
braces ({}), 17
break points, 219
browser (debugging tool), 216
browser mode, 217
bytes, 42
bzfile function, 206
bzip2 compression, 206

C

c (concatenate) function, 25, 38, 150
calculator, 6
calling environment, 103
capitalization, 9
case sensitivity, 9
categorical information, 50
character atomic vectors, 38, 41
character strings
    assembling, 41

preventing factoring of, 203
vs. factors, 50
vs. R objects, 41
class attribute, 47
classes, creating new, 153
    (see also S3 class system)
closures, 108
code
    code comments, 138
    compilation of, 5, 179
    creating drafts of with R scripts, 20
    creating speed in (see vectorized code)
    readability of, 17
    strategy for coding, 120, 140
coercion, 51
colon operator (:), 5, 7
columns, removing from data frames, 78
comma-separated-values file (.csv), 58, 201, 209
command line
    definition of, 4
    downloading R packages from, 24
    error messages at, 6
    keyword help page search, 31
commands
    canceling, 6
    definition of, 4
    repeating with replicate, 27
commenting symbol, 6
comparisons, 42, 82
complex atomic vectors, 38, 42
concatenate (c) function, 25, 38, 150
copy/paste, from Excel spreadsheets, 209
CRAN webpage, 24, 189, 194
createSheet, 210
ctrl + c, 6

D

data
    boolean, 42
    coercion of data types, 51
    examining imported, 59
    importing from the internet, 59
    loading/saving
        data sets in base R, 199
        Excel spreadsheets, 208–211
        in databases, 212
        in R objects, 57, 61
        loading files from other programs, 211
        plain text files, 200–207

R files, 207
   working directory, 200
   multiple types stored in R objects, 51
   object types available for, 61
   storing as logical vectors, 42
   storing as one-dimensional sets, 53
   storing as two-dimensional lists, 55
   storing bytes of, 42
   storing categorical information, 50
   storing multiple types of, 53
   types stored in R objects, 38
data frames
   attributes and, 142
   benefits of, 55
   examining imported data with, 59
   removing columns from, 78
   selecting columns from, 68, 73
data science
   benefits of programming skills for, 1
   core skill sets needed for, 187
   foundation of, 34
   problem types faced in, 186
data sets
   attaching, 76
   examining, 60
   in base R, 199
   loading with RStudio import wizard, 58
   manual entry of, 57
   multiple data types in, 51
   using, 199
databases, connecting to, 212
dates/times, 48
DBI package, 212
debug function, 220
debugging tools
   break points, 219
   browser, 216
   debug, 220
   for S3 class system, 154
   recover, 222
   trace, 221
   traceback, 213
debugonce, 221
devtools R package
   installing R from alternate sources, 195
dimensions (dim) attribute, 45
dispatch, of methods, 150
dollar sign ($), 73, 97
double atomic vectors, 38, 39, 48
double equals sign (==), 81
double hashtag character (##), 6
drop = FALSE argument, 68
dynamic programing languages, 5

E

element-wise execution, 10, 81, 86, 173
else statements, 125–132
endCol argument, 210
endRow argument, 210
environment function, 97
Environment pane, 8, 9
environments
   accessing objects in specific, 97
   active, 97, 99
   assignment and, 99
   calling environment, 103
   closures, 108
   displaying with parenvs function, 94
   empty, 96
   evaluation, 99–108
   global, 96
   helper functions for, 95
   hierarchical arrangement of, 93
   looking up parent environment, 96
   origin, 101
   runtime environments, 99
   saving objects into particular, 97
   scoping rules, 98
   viewing objects saved in, 96
   visualizing structure of, 95
equality operator (==), 81
equals sign (=), 81
errors/error messages
   argument values, 19
   at command line interface, 6
   floating point errors, 40
   in object calls, 98
   positive/negative integer indexing, 68
   strategy for dealing with, 140
   when creating character strings, 41
   when naming arguments, 13
   with if statements, 123
ERSI ArcGIS format, 211
evaluation, 99–108
Excel spreadsheets
   copying/pasting data from, 209
   data frames, 55
   exporting data, 208

reading, 209
writing to, 210
XLConnect package, 209, 211
expand.grid function and, 159–164
expected values, 157
expressions, 123
extract function, 22, 138

F

F (FALSE), 42, 69, 80, 123
factorial function, 12
factoring, preventing, 56, 59, 203
factors, 50
FALSE (F), 42, 69, 80, 123
files
    fixed-width files (.fwf), 204
    HTML links in, 205
    listing in working directory, 200
    loading/saving from other programs, 211
    missing information in, 201
    plain text files, 200
        (see also plain text files)
    R files
        opening/saving, 207
        vs. plain text files, 208
    RData files
        opening, 207
        saving, 207
        vs. RDS files, 207
    RDS files
        opening, 207
        saving, 207
        vs. RData files, 207
    RStudio file import wizard, 58
fixed-width files (.fwf), 204
floating point arithmetic, 40
floating point errors, 40
for loops, 165–170, 180
foreign package, 211
function constructor, 17
functions
    accessor functions, 96
    all, 85, 123
    any, 85, 123
    arguments for, 12
    array, 46
    as verbs, 34
    as.character, 50
    as.environment, 95

assign, 97
attr, 143
base R collection of, 193
basics of, 16
body of, 18
environment, 97
expand.grid function, 159–164
extract function, 22, 138
factorial, 12
generic, 147
head, 60
help pages for, 29
helper functions, 44, 95, 142
install.packages, 193
is.na, 90
levels, 142
ls, 9, 96
ls.str, 96
matrix function, 46
mean, 90
multiple arguments for, 13, 19
order of execution, 12
parenvs, 94
parts of, 17
print, 119, 141, 147
replicate, 27
round, 12
row.names, 142
R_EmptyEnv, 96
R_GlobalEnv, 96
sample function, 12, 29
sample with replacement, 14
show_env, 100
storage of, 99
sum function, 15
Sys.time, 174
system.time, 174
tail, 60
temporary storage of, 99
unique, 129
UseMethod, 148, 150
using, 12
writing your own, 17

G

generic functions, 147
getwd(), 61, 200
ggplot2 package, 23
global environment, 96

gnu zip (gzip) compression, 206
graphs
    histograms, 26
    qplot function for, 23
    scatterplots, 25
gzfile function, 206

H

hard bracket, single ([), 26, 65, 74
hard brackets, double ([[), 73
hashtag character (#), 6
head function, 60
head(deck), 59
header argument, 201
headers, 58
help
    help pages, 29
    user communities, 33
helper functions
    array, 46
    as.environment, 95
    for environments, 95
    levels, 142
    matrix, 46
    purpose of, 44
    row.names, 142
histograms, 26
HTML links, in plain text files, 205
human readable code, 5

I

i, 42
if statements, 122
ij notation, 66
independent random samples, 15
indexes
    0 vs. 1 as starting point, 68
    blank spaces, 69
    logical values, 69
    name, 70
    negative integers, 68
    positive integers, 66
    writing, 65
    zero, 69
infinite recursion errors, 215
install.packages function, 24, 193
integer atomic vectors, 38, 40

integers
    negative, 68
    positive, 66
is.na function, 90
is.vector(), 38

K

keywords, 31

L

L, 40
levels function, 142
library() command, 24, 194
linear algebra, 66
list.files(), 200
lists, 53, 73
load function, 207
loadWorkbook, 209
logical atomic vectors, 38, 42
logical operators, 81
logical subsetting, 80
logical tests, 80, 123, 173
logical value indexing, 69
lookup tables, 132–137
loops
    benefits/drawbacks of, 171
    expand.grid function and, 159–164
    expected values and, 157
    for loops, 165–170
    repeat, 171
    saving output of, 171
    vs. vectorized code, 185
    while loops, 170
ls function, 9, 96
ls.str function, 96

M

Mac R GUI, 191
machine readable code, 5
Matlab format, 211
matrix function, 46
matrix multiplication, 10
mean function, 90
messages, displaying, 119, 147
    (see also errors/error messages)
metadata, 43
methods, 148–152

minitab format, 212
missing information
　in plain text files, 201
　managing, 89
multiple data types, storing, 53
multiplication
　inner/outer operators for, 11
　matrix vs. element-wise execution, 10

N

NA character, 89
na.rm argument, 90
na.strings argument, 201
names/naming
　attributes, 144
　data frames, 56
　displaying previously used, 9
　name indexing, 70
　names attribute, 44
　of arguments, 13
　of methods, 150
　rules for R objects, 8
nrow argument, 202
NULL, 43, 78
numbers
　creating a sequence of, 5, 7
　manipulating sets of, 9
　returning a vector of, 7
　returning vectors with c, 25
　saving as R objects, 7
　storing as double vectors, 39
　storing as integer vectors, 40
　storing complex, 42
　storing in n-dimensional arrays, 46
　storing in two-dimensional arrays, 46
　vs. strings, 41
numerics (see double atomic vectors)

O

Object-Oriented Programming (OOP), 155
objects
　definition of, 7
　　(see also R objects)
　functions as, 99
　temporary, 99
online data, 59
origin environment, 101

P

packages, 193
　(see also R packages)
　DBI, 212
　foreign, 211
　ggplot2, 23
　RODBC, 212
　XLConnect, 209, 211
parallel cases subtask, 121
paren.env, 96
parent environment, 94
parentheses), 26
parenvs function, 94
plain text files
　benefits of, 200, 208
　compressing, 206
　header argument, 201
　HTML links in, 205
　loading, 58, 201, 203
　na.strings argument, 201
　nrow argument, 202
　read.fwf, 204
　saving, 205
　sep argument, 201
　skip argument, 202
　stringAsFactors argument, 203
　table structure in, 201
　vs. R files, 208
playing cards project
　building entire deck, 56
　changing card point values, 77
　creating a copy of, 77
　creating card deck, 38
　dealing cards, 70, 104
　downloading deck data frame, 57
　finished deck appearance, 37
　locating individual cards, 86
　overview of, 35
　saving card names, 43
　saving data, 61
　shuffling cards, 71, 108
　storing cards in runtime environment, 108
　storing cards with lists, 54
positive integers, 66
POSIXct class, 48
POSIXt class, 48
print function, 119, 141, 147
print.factor method, 149
print.POSIXct method, 149

probability, 160
programs
    code comments, 138
    displaying messages in, 119, 147
    else statements, 125–132
    if statements, 122
    improving performance of (see vectorized code)
    lookup tables, 132–137
    parallel cases subtask, 121
    saving, 119
    sequential steps subtask, 120
    strategy for coding, 120, 128, 140
    wrapping into functions, 138
pryr R package
    parenvs function, 94

Q

qplot function
    downloading/installing, 23
    histogram creation with, 26
    scatterplot creation with, 25
question mark (?), 29
quick plots, 25
quote mark ("), 41

R

R
    32- vs. 64-bit versions, 190
    base R, 193
    browser mode, 217
    building eloquent statements in, 34
    building from source files, 190
    downloading/installing, 189
    dynamic programming capabilities, 5
    help pages, 29
    installing from binaries, 190
    opening, 192
    updating, 197
    user communities, 33
    user interface (see RStudio)
    using with RStudio, 191
    using with Unix, 191
    vs. R language, 5
R expressions, 123
R files
    opening/saving, 207
    vs. plain text files, 208

R library, displaying current packages in, 194
R notation system
    blank space indexing, 69
    changing values in place, 77
    dollar sign ($), 73, 97
    extracting values from R objects, 65
    hard brackets ([]), 73
    hard brackets, double ([[]]), 73
    logical values indexing, 69
    name indexing, 70
    negative integer indexing, 68
    positive integer indexing, 66
    zero indexing, 69
R objects
    accessing, 9
    accessing in specific environments, 97
    arrays, 46
    as nouns, 34
    atomic vectors
        characters, 41
        complex, 42
        creating with c, 38
        discovering type of, 39
        doubles, 39
        integers, 40
        logicals, 42
        raw, 42
        testing for, 38
        types of, 38
    attributes
        as metadata, 43
        dim, 45
        displaying, 43, 142
        names, 44, 144
    class, 47
    coercion, 51
    creating, 8
    creating empty, 69
    creating NULL objects, 43
    data frames, 55, 59, 68, 73
    dates/times, 48
    element-wise execution in, 10
    extracting values from (see R notation)
    factors, 50
    lists, 53, 73
    loading data, 57
    manipulation of (see environments)
    matrices, 46
    modifying values in place, 77

naming, 8
saving into particular environment, 97
scoping rules and, 98
types available, 61
updating, 97
vector recycling in, 10
viewing saved in environments, 96
vs. character strings, 41
R packages
benefits of, 193, 195
displaying currently installed, 194
installing, 193
installing from alternate sources, 195
installing from command line, 24
installing from mirrors, 194
installing multiple, 194
list of available, 195
loading, 194
selecting, 195
updating, 197
using, 24
R scripts, 20, 119
R.matlab library, 211
random samples, 15
raw atomic vectors, 38, 42
RData files
opening, 207
saving, 207
vs. RDS files, 207
RDS files
opening, 207
saving, 207
vs. RData files, 207
read.csv, 203
read.csv2, 203
read.delim, 203
read.delim2, 203
read.dta function, 212
read.fwf, 204
read.mtp function, 212
read.shapefile function, 211
read.spss function, 212
read.ssd function, 212
read.systat function, 212
read.table, 201, 203, 209
read.xport function, 212
readMat function, 211
readRDS function, 207
readWorksheet, 209

recover function, 222
recursion errors, 215
repeat loops, 171
replicate function, 27
RODBC package, 212
round function, 12
row.names = FALSE, 61
row.names function, 142
RStudio
changing working directory in, 200
command line interface, 4
data viewer, 59
downloading, 191
Environment pane, 8, 9, 96
Extract Function option, 138
file import wizard, 58
GUIs for Windows and Mac, 191
Help tab, 29
IDE for R, 192
opening R scripts in, 119
recognizing infinite recursion errors, 215
Show Traceback option, 215
Traceback pane, 217
updating, 197
rule of probability, 160
runtime environments, 99
R_EmptyEnv, 96
R_GlobalEnv, 96

S

S3 class system
alternatives to, 154
attributes, 142–147
classes, 153
debugging, 154
example of, 141
generic functions, 147
methods, 148–152
origins of, 150
overview of, 155
sample function, 12, 14, 29, 117
SAS (permanent data set) format, 212
SAS (XPORT format), 212
save function, 207
saveRDS function, 207
saveWorkbook, 211
scatterplots, 25
scoping rules, 98
scripts, 20, 119

sep argument, 201
sequential steps subtask, 120
setwd(), 200
shapefiles library, 211
sheet argument, 209
show_env function, 100
skip argument, 202
slot machine project
    calculating payout rate, 157
    calculating prize value, 161
    calculating prizes, 119, 121
    creating play function, 119, 120
    displaying prizes, 141
    generating/selecting symbols, 117
    Manitoba Video Lottery payout scheme, 118
    saving your program, 119
    subtask skeleton, 126
    symbol test subtask, 127
source files, building R from, 190
special characters, in names, 8
speed (see vectorized code)
spreadsheets (see Excel spreadsheets)
SPSS format, 212
Stack Overflow website, 33
startCol argument, 210
startRow argument, 210
Stata format, 212
str function, 56
stringAsFactors argument, 56, 203
strings
    assembling, 41
    vs. numbers, 41
    vs. R objects, 41
    (see also character strings)
subtasks
    parallel cases, 121
    sequential steps, 120
    skeleton representation of, 126
    strategy for coding, 128, 140
sum function, 15
symbols, in names, 8
sys.time(), 48, 174
Systat format, 212
system.time function, 174

T

T (TRUE), 42, 69, 80, 123
tail function, 60
tasks, repeating (see loops)

text, storing as character vectors, 41
trace function, 221
traceback tool, 213
trees
    converting lookup tables to, 137
    else if statements, 131
    else trees, 125
    inefficiency of, 135
    vs. lookup tables, 136
TRUE (T), 42, 69, 80, 123
typeof(), 39

U

unclass, 50
unique function, 129
Universal Coordinated Time Zone (UTC), 49
Unix
    using R with, 191
    working directory, 200
update.packages command, 197
UseMethod function, 148, 150

V

values
    average value, 158
    boolean operators for subsetting, 85
    capturing all possible combinations of, 159
    changing in place, 77
    creating new, 78
    expected values, 157
    logical tests for subsetting, 80
    managing missing, 89
    replacing, 78
    selecting from data frames/lists, 73
    selecting from R objects, 65
    selecting multiple, 66
vectorized code
    benefits of, 186
    example of, 173
    for loops and, 180
    using, 181
    vs. loops, 185
    writing, 175–180
vectors
    boolean operators and, 86
    coercion in, 51
    element-wise execution and, 10
    in histograms, 26

in scatterplot creation, 25
multiple data types in, 51
returning with : operator, 7
vector recycling, 10

W

weighted dice project
    access virtual dice, 9
    adding the dice, 15
    biasing the rolls, 28
    checking dice accuracy, 27
    creating R object, 8
    creating virtual dice, 7
    frequency of fair dice combinations, 27
    re-rolling the dice, 16
    rolling the dice, 12
    simulating dice pairs, 14
    simulating repeated rolls, 28
while loops, 170
Windows R GUI, 191

working directory
    listing files in, 200
    locating/changing, 61, 200
    moving, 200
write.csv, 61, 205
write.csv2, 205
write.table, 205
writeWorksheet, 210
writeWorksheetToFile, 211

X

XLConnect package, 209, 211
xor operator, 85
xz compression, 206
xzfile function, 206

Z

zero indexing, 69

About the Author

Garrett Grolemund is a statistician, teacher, and R developer who works for RStudio. He sees data science as a largely untapped fountain of value for both industry and academia. Garrett received his PhD at Rice University in Hadley Wickham's lab, where his research traced the origins of data science as a cognitive process and identified how attentional and epistemological concerns guide every data analysis.

Garrett is passionate about helping people avoid the frustration and unnecessary learning he went through while mastering data science. Even before he finished his dissertation, he started leading corporate trainings in R and data analysis for Revolution Analytics. He's taught at Google, eBay, Roche, and many other companies, and is developing a training curriculum for RStudio that will make useful know-how even more accessible.

Outside of teaching, Garrett spends time doing clinical trials research, legal research, and financial analysis. He also develops R software, has coauthored the lubridate R package—which provides methods to parse, manipulate, and do arithmetic with date-times—and wrote the ggsubplot package, which extends the ggplot2 package.

Colophon

The animal on the cover of *Hands-On Programming with R* is an orange-winged Amazon parrot (*Amazona amazonica*). *Loros guaros*, as the birds are known locally, reside year-round in the humid tropics east of the Andes in South America, from Colombia and Venezuela in the north to Central Brazil in the south.

Orange-winged Amazons are both voluble and sociable. Quiet only when feeding, these birds roost communally in tree tops with as many as a thousand counterparts. Males of the species, who regurgitate food for a female partner while she incubates eggs and regurgitates food in turn to feed the brood, keep quarters near the nest during the day but return to the flock at night. They are often seen on morning and evening flights from tree-top roosts to feeding sites or nests in tree cavities.

The orange-winged variety displays the generally green plumage typical to other Amazon parrots, from whom it is distinguished by its orange speculum feathers. The forehead of the orange-winged Amazon is, like that of the blue-fronted Amazon, covered in blue feathers which give way to yellow on the crown and cheeks. The horn color at the base of the orange-winged Amazon's beak becomes a dark gray at the tip. Males and females exhibit no stable difference in appearance.

Many of the animals on O'Reilly covers are endangered; all of them are important to the world. To learn more about how you can help, go to animals.oreilly.com.

The cover image is from *Meyers Kleines Lexicon*. The cover fonts are URW Typewriter and Guardian Sans. The text font is Adobe Minion Pro; the heading font is Adobe Myriad Condensed; and the code font is Dalton Maag's Ubuntu Mono.

Have it your way.

O'Reilly eBooks

- Lifetime access to the book when you buy through oreilly.com
- Provided in up to four, DRM-free file formats, for use on the devices of your choice: PDF, .epub, Kindle-compatible .mobi, and Android .apk
- Fully searchable, with copy-and-paste, and print functionality
- We also alert you when we've updated the files with corrections and additions.

oreilly.com/ebooks/

Safari Books Online

- Access the contents and quickly search over 7000 books on technology, business, and certification guides
- Learn from expert video tutorials, and explore thousands of hours of video on technology and design topics
- Download whole books or chapters in PDF format, at no extra cost, to print or read on the go
- Early access to books as they're being written
- Interact directly with authors of upcoming books
- Save up to 35% on O'Reilly print books

See the complete Safari Library at safaribooksonline.com

©2014 O'Reilly Media, Inc. O'Reilly logo is a registered trademark of O'Reilly Media, Inc. 14373

Get even more for your money.

Join the O'Reilly Community, and register the O'Reilly books you own. It's free, and you'll get:

- $4.99 ebook upgrade offer
- 40% upgrade offer on O'Reilly print books
- Membership discounts on books and events
- Free lifetime updates to ebooks and videos
- Multiple ebook formats, DRM FREE
- Participation in the O'Reilly community
- Newsletters
- Account management
- 100% Satisfaction Guarantee

Signing up is easy:

1. Go to: oreilly.com/go/register
2. Create an O'Reilly login.
3. Provide your address.
4. Register your books.

Note: English-language books only

To order books online:
oreilly.com/store

For questions about products or an order:
orders@oreilly.com

To sign up to get topic-specific email announcements and/or news about upcoming books, conferences, special offers, and new technologies:
elists@oreilly.com

For technical questions about book content:
booktech@oreilly.com

To submit new book proposals to our editors:
proposals@oreilly.com

O'Reilly books are available in multiple DRM-free ebook formats. For more information:
oreilly.com/ebooks